J. M. TEDDER
University of St. Andrews

A. NECHVATAL and **A. W. MURRAY**
University of Dundee

J. CARNDUFF
University of Glasgow

Basic Organic Chemistry

Part 3

1970
John Wiley and Sons
London   New York   Sydney   Toronto

Library of Congress catalog card number: 66-17112

ISBN 0 471 85013 6

Printed in Great Britain
By Unwin Brothers Limited
The Gresham Press, Old Woking, Surrey, England

Basic
Organic
Chemistry
Part 3

Preface to Part 3

This volume, like the previous two, is largely based on lectures given at the Universities of St. Andrews and Dundee. The present chapters now correspond to third year lectures (2nd year in England and Wales) and the various topics are necessarily dealt with in greater depth. We feel that no one person can have sufficient depth of knowledge or adequate perspective in the various branches of organic chemistry at this level to enable them to write in a truly authoritative fashion. We have made no real attempt at uniformity in style of writing, and hence this may change from chapter to chapter and, on occasion, even from section to section of the same chapter. Though the style of writing may change, the objectives remain identical throughout. As in Parts 1 and 2 we have attempted to develop the current ideas of organic chemistry and to show that the vast range of reactions undergone by the many different classes of compound can be fitted into the same general mechanistic picture.

Chapters 1 and 2 deal with aromaticity first in carbocyclic and then in heterocyclic compounds. Chapters 3 and 4 deal with alicyclic and saturated heterocyclic compounds. The emphasis here is on molecular shape and its effect on chemical reactions. Chapter 5 introduces the organic chemistry of second-row elements, with the emphasis very much on basic principles rather than experimental detail. Chapter 6 is a discussion of how reaction mechanisms can be established, and Chapter 7 introduces simple quantum mechanics into organic chemistry.

Like the previous two volumes this book assumes a continually increasing knowledge of the other branches of chemistry and of laboratory methods. There has been no attempt to cover the thermodynamics or the basic quantum mechanics necessary for a full understanding of the later chapters. Equally important is the assumption that by the time students use this volume they will have had extensive experience in the interpretation of spectra. Only the barest outlines were covered in Part 2 as we believe that the necessary experience must be gained in the laboratory.

The purpose of these books is to provide the student with readable handbooks in which the basic principles of organic chemistry are presented in a mechanistic scheme. They are not compendia of facts, and the students must have access to conventional textbooks and, more especially, specialist monographs dealing with particular aspects of the subject; for this reason a brief bibliography is included at the end of each chapter.

Part 1 is an introduction which can be used by itself, and Parts 2 and 3 belong together and are intended to cover the main concepts of organic chemistry, developed from the basic ideas introduced in Part 1. The three volumes together are intended to provide an undergraduate text from which only a treatment of natural products is omitted. This will be covered in Part 4 where it will be shown that the ideas developed in the first three parts can be applied to the complex carbon compounds occurring in Nature.

Acknowledgements

We are very grateful to the many colleagues who have given us advice and who have read through chapters for us. Without their help many errors might have been included; those that may remain are, of course, entirely the responsibility of the authors. We would particularly like to acknowledge the help of Professor P. L. Pauson (Strathclyde), D. M. G. Lloyd (St. Andrews), and Dr. K. M. Watson (Dundee) in Chapter 1; of Professor Pauson, Dr. Watson, and Dr. D. H. Reid (St. Andrews) in Chapter 2; of Dr. K. H. Overton (Glasgow) and Dr. F. G. Riddell (Stirling) in Chapters 3 and 4; of Dr. J. A. Miller (Dundee), Professor J. I. G. Cadogan (Edinburgh), and Professor S. Trippett (Leicester) in Chapter 5; of Professor V. Gold (King's College, London), Professor P. A. H. Wyatt (St. Andrews), and Dr. A. R. Butler (St. Andrews) in Chapter 6; of Dr. E. T. Stewart (Dundee) and Dr. C. Thomson (St. Andrews) in Chapter 7. We are even more indebted to Professor T. S. Stevens (Strathclyde), Dr. R. S. Cahn, and Dr. R. Brettle (Sheffield) for reading the whole manuscript. Above all, our thanks go to our students whose response to our lectures and discussion in tutorials has made the book possible.

J.M.T.
A.N.
A.W.M.
J.C.

UNITS

We have written this book at a time when units and the symbols for them are being changed. A new international system (SI units) is being gradually adopted. Because we are in a period of transition we have in most cases put the data in the previously accepted units after the SI units (e.g. 'the resonance energy of benzene is estimated as approximately 148 kJ mole^{-1}, i.e. 36 kcal mole^{-1}'). The units we are using, and their relationship to other common units, are tabulated below.

Thermodynamic Data

Energies in joules, J (1 calorie = 4.184 J).

Heats of reaction, ΔH, in kilojoules mole^{-1} (kJ mole^{-1}).

Entropies, S, in joules degree^{-1}.

Entropies of reaction, ΔS, in joules degree^{-1} mole^{-1} (J K^{-1} mole^{-1}).

Ultraviolet Spectra

Wavelength in nanometres, nm (10^{-9} m = 1 nm).

Molar extinction coefficient, ϵ, in litres mole^{-1} cm^{-1}, i.e. 10^3 cm^2 mole^{-1}. (Units are often omitted, but this could be ambiguous since the SI units are m^2 mole^{-1}.)

Infrared Spectra

Wavenumber in cm^{-1}.

The position of an infrared band in a spectrum may be denoted by its wavelength (λ_{max} in μm) or by its wavenumber (ν_{max} in cm^{-1}). [Frequency in s^{-1} = 3×10^{10} wavenumber in cm^{-1}.]

Nuclear Magnetic Resonance Spectra

Chemical shifts and coupling constants in hertz (Hz = s^{-1}).

Bond Lengths

Lengths in ångstroms (1 Å = 10^{-1} nm).

Dipole Moments

Dipole moments in debyes [D = 10^{-18} cm$^{5/2}$ g$^{1/2}$ sec^{-1} (cf. p. 45 of Part 2) = 3.336×10^{-30} Cm (C = coulomb)].

Contents

ix

CHAPTER 1

Aromaticity

In Chapter 13 of Part 1 we saw that there is no such compound as cyclohexatriene with alternate double and single bonds; instead the molecule C_6H_6, called benzene, has its atoms arranged hexagonally, with the carbon–carbon bond lengths intermediate between those of a double and a single bond. The striking feature of the chemistry of benzene is its low reactivity compared with the reactivity that we would expect for the hypothetical cyclohexatriene. Benzene does not react immediately with bromine, it is hydrogenated only with difficulty, it is unaffected by ordinary oxidizing agents, and, although powerful electrophiles will add to benzene, the overall reaction is an 'addition-with-elimination'. Originally, the reactions of benzene with electrophiles were called substitution reactions, a name that correctly describes the overall consequences of the reaction. We now know that the initial stage of the reaction is the addition of the electrophile in exactly the same way as the electrophile would add to an ethylenic double bond. The difference is that the carbonium ion generated by the addition of an electrophile to an olefin adds an anion (thus 'saturating' the double bond) whereas the benzene adduct (called in Part 2 the 'Wheland intermediate') ejects a proton, thus regenerating the resonance-stabilized benzene ring. Thus, we saw in Part 1 that benzene differs from the hypothetical cyclohexatriene and from linear polyenes because of the great stability of the resonance-stabilized ring. We saw that the 'addition-with-elimination' reactions which result in substitution are characteristic of benzene, and we regarded benzene as the archetype of a group of cyclic molecules called *aromatic compounds*, which undergo the same type of reaction.

In Part 2 we looked at this problem from a different angle. We

1

saw that butadiene could be considered to be formed from four sp^2 hybridized carbon atoms (Figure 1.1) such that the non-hybridized p_y orbitals on each atom could be combined to give four π molecular orbitals, two bonding and two antibonding (Figure 1.2). We saw that this delocalization of electrons in π orbitals resulted in the lowering of energy so that the total π electron energy of butadiene is slightly less than twice that of an isolated olefinic bond.

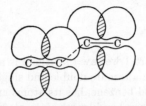

Figure 1.1. Diagram representing overlapping $2p_y$ atomic orbitals in butadiene.

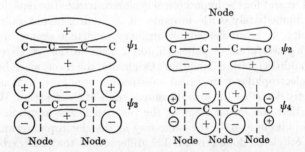

Figure 1.2. Four π molecular orbitals formed by combining the four $2p_y$ orbitals in butadiene. Energy increasing in the order
$$\Psi_1 < \Psi_2 < \Psi_3 < \Psi_4.$$

The energy levels of the four π molecular orbitals of butadiene occur in pairs; thus the energy required to promote an electron to ψ_3 is exactly equal to the energy released by placing an electron in ψ_2, and likewise the energy required to promote an electron to ψ_4 is equal to the energy released when an electron is placed in ψ_1. In a similar way the π molecular orbitals of hexa-1,3,5-triene and octa-1,3,5,7-tetraene have symmetrically arranged energy levels (see Figure 1.3).

In Part 2 we depicted benzene as formed from six sp^2 hybridized carbon atoms arranged in a regular hexagon to provide maximum overlap of the hybridized orbitals (Figure 1.4). We then combined the six non-hybridized $2p_y$ atomic orbitals of the carbon atoms to form six molecular π orbitals (Figure 1.5).

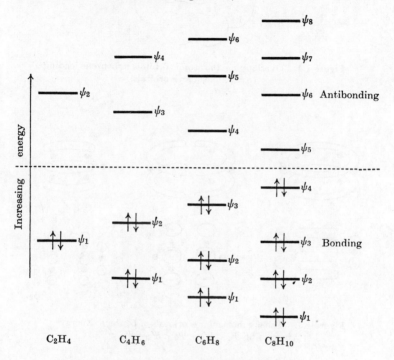

Figure 1.3. Orbital energies for linear polyenes.

The energy levels of the six π orbitals of benzene are arranged symmetrically above and below the non-bonding level, as with the linear polyenes, but we have an additional symmetry, namely, that ψ_2 and ψ_3 are of equal energy, i.e. degenerate. Our energy level diagram for benzene is shown in Figure 1.6. The lowering of energy associated with the electron delocalization in the π orbitals of benzene is very considerable indeed; thus, the resonance energy (the energy gained by electron delocalization) of buta-1,3-diene is approximately

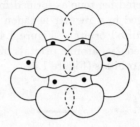

Figure 1.4. Overlapping $2p_y$ atomic orbitals in benzene leading to six molecular π orbitals.

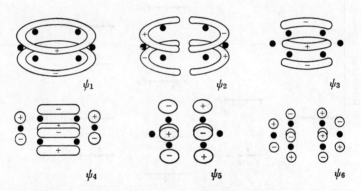

Figure 1.5. The six molecular π orbitals of benzene. Energy increasing $\Psi_1 < \Psi_2 = \Psi_3 < \Psi_4 = \Psi_5 < \Psi_6$.

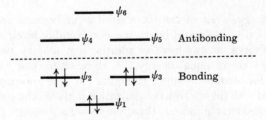

Figure 1.6. Orbital energies for benzene.

14.6 kJ mole^{-1}, i.e. 3.5 kcal mole^{-1}, whereas the resonance energy of benzene is estimated as approximately 148 kJ mole^{-1}, i.e. 36 kcal mole^{-1}.

In Part 1 we considered the possible existence of other cyclic molecules with a general formula C_nH_n, and we concluded that the failure to prepare cyclobutadiene could, at least in part, be attributed to the energy needed to bend the bonds in making this molecule. Similarly, we concluded that one of the reasons why cyclooctatetraene was a puckered molecule, behaving like an ordinary conjugated diene, was that more energy would be required to bend the ethylenic bonds to make it a planar molecule and enable resonance to occur than would be gained by the resultant resonance. If we now look at this problem in terms of the pictures we developed in Part 2 we can see that we have two problems to consider. If we suppose that we are forming cyclic compounds from sp^2 hybridized carbon atoms, we must first consider how effective the overlap of the sp^2 hybridized atomic orbitals, which are to form the σ bonds of our cyclic molecule, will be. In the case of cyclobutadiene considered above we can see that if the four carbon atoms are placed at the corners of the square the overlap between the sp^2 hybrids formed from these atoms will be rather poor. We could, of course, improve the overlap by changing the hybridization, but our new hybrid orbitals would be of higher energy than the sp^2 hybrid orbitals, so that although the overlap would be better the overall energy situation would not. Putting it in another way, we see that our arguments about bond distortion in Part 1 are entirely consistent with the picture we built up in Part 2 and, leaving other considerations aside, we should expect planar ring systems of 3, 4, 8, or 9 carbon atoms to be less stable than those of 5, 6, or 7 atoms. The overlap of sp^2 hybrids in a five-membered or a seven-membered ring is not as good as that in a six-membered ring but it is still quite favourable. Having considered the σ orbitals of our cyclic compounds we must now consider what their π molecular orbitals will be.

We have seen that π orbitals of a conjugated polyene occur in pairs, each bonding orbital having an energy of say $-x$ J mole^{-1} and a corresponding antibonding orbital of energy $+x$ J mole^{-1}. When we come to consider a cyclic polyene there is further symmetry in the energy levels of the π orbitals. If the molecule contains an even number of carbon atoms in the ring, the lowest-energy bonding orbital and the highest-energy antibonding orbital will be non-degenerate, but all the remaining orbitals will occur in pairs of degenerate orbitals. As we have seen with benzene, ψ_1 and ψ_6 are non-degenerate while ψ_2 and ψ_3 are degenerate and have a bonding energy corresponding to the antibonding energy of ψ_4 and ψ_5 which are also a degenerate pair. If the cyclic compound has an odd number of carbon atoms in the ring, then the lowest-energy bonding orbital is non-degenerate but all the remaining orbitals occur in degenerate pairs. We can thus depict the energy levels of cyclic compounds in which all the bond lengths are equal as shown in Figure 1.7.

Neutral C_3, C_5, and C_7 compounds must be free radicals, since each contains an odd number of electrons and thus one electron must be left unpaired. Neutral C_4, C_6, and C_8 compounds have an even number of electrons and, in terms of simple resonance theory, we can write two Kekulé structures for cyclobutadiene and planar cyclooctatetraene, exactly analogous to the two Kekulé structures for benzene. Notice that when we come to assign the π electrons to their orbitals we have to place two electrons with parallel spins in a pair of degenerate orbitals if we are to obey Hund's rules. Thus, apparently the ground state of square cyclobutadiene and octagonal planar cyclooctatetraene would in both cases be a triplet or a diradical. More detailed theories suggest that delocalization of the π electrons in these molecules might have a destabilizing effect (so called 'anti-aromaticity'). In other words, these theories predict that such molecules will exist as polyenes with alternate single and double bonds. We will consider below the attempts that have been made to make cyclobutadiene, and see to what extent the predictions of simple molecular-orbital theory have been borne out. However, when we come to consider benzene we find that all six π electrons can be placed in pairs with opposing spins in bonding orbitals. We now see that the uniqueness of benzene compared with cyclobutadiene or cyclooctatetraene is more than simply a question of bond angles. If we look at the π electrons of the cyclopentadienyl radical we can see that the addition of one electron to the half-filled bonding orbital gives the molecule three filled π orbitals analogous to those of benzene. We should thus expect the cyclopentadienyl anion to be particularly stable and to show *some* of the characteristics of benzene. In a similar way the cycloheptatrienyl radical, C_7H_7, can lose an electron from its half-filled antibonding orbital, again leaving three filled π orbitals

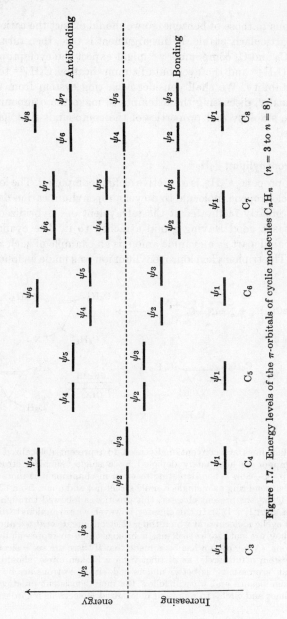

Figure 1.7. Energy levels of the π-orbitals of cyclic molecules C_nH_n $(n = 3$ to $n = 8)$.

analogous to those of benzene, so we should expect the cation $C_7H_7^+$ **to** be particularly stable. If this argument is true, then turning back to the C_3 and C_4 compounds we might expect the cyclopropenylium cation $C_3H_3^+$ and the cyclobutadienium dication $C_4H_4^{2+}$ to exhibit some stability. We shall consider each ring system from C_3 to C_8 individually, discussing the attempts to make the compound C_nH_n and the stability and properties of the compounds that have been prepared.

Cyclopropenylium, $C_3H_3^+$

Cyclopropene, C_3H_4, is a reactive olefinic compound. The ionization of a cyclopropene molecule to a cyclopropenylium cation derivative will be greatly facilitated if the tetravalent sp^3 hybridized carbon atom has a good leaving group attached to it. The cyano group, which can depart as a cyanide anion, is an example of such a leaving group. The triphenylcyclopropenylium ion was made as follows:*

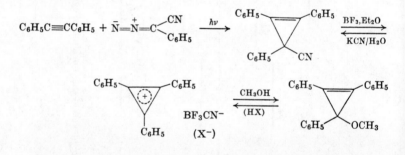

* Broken lines are conventionally used to represent delocalized electrons which cannot be adequately depicted by a single canonical structure (cf. Part 1, p. 85—86). When depicting reaction mechanisms it is more useful to draw a benzene ring as a single Kekulé structure, and apart from Chapter 13 of Part 1 and the present chapter, this practice is followed throughout these books (cf. Part 1, p. 133). In this chapter, however, we are anxious to distinguish between cyclic molecules in which the π electrons are delocalized and those in which they are not, so we will use a broken circle to represent delocalized π electrons. (A suggestion has been made that if there are six π electrons in a cyclic system a full circle, as distinct from a broken circle, should be used. Now that 'aromaticity' is better understood, six π electrons are no longer of special significance and we will follow the more consistent practice of using broken lines and circles to represent delocalized π electrons regardless of their number.)

Photolysis of the cyano-phenyldiazomethane yields the corresponding carbene which adds to diphenylacetylene to yield the cyano-tri-phenylcyclopropene. The boron atom in boron trifluoride has an incomplete outer shell of electrons, and when dissolved in ether the boron atom accepts two electrons from the ether oxygen atom $[F_3B^- {}^+O(C_2H_5)_2]$. If a better nucleophile is available the boron atom will preferentially accept electrons from it and release the ether molecule. Thus, in the present experiment, if the cyano group shows any tendency to ionize, forming the cyanide anion, it will immediately become bound to the boron atom of the boron tri-fluoride.

Cyclopropenylium salts are quite stable; the ring atoms are arranged at the corners of an equilateral triangle and there is a π electron cloud above and below the plane of the ring exactly as in benzene. The stability of the cation is not dependent on conjugation with the three phenyl groups in the compound we have depicted. The corresponding tripropylcyclopropenylium ion is just as stable. The question arises, is it sensible to describe this system as aromatic? The answer is surely, no. The cyclopropenylium ion has some of the properties of benzene. It is planar and it has delocalized π electrons. The cyclopropenylium ion has an electronic arrangement analogous to that of benzene but none of the chemical properties of benzene; it is, in fact, a typical stable carbonium ion. It reacts with nucleophiles in the usual way, and we have depicted the re-formation of the cyano-triphenylcyclopropene from the cation and also the reaction of the same cation with methanol to yield a corresponding ether. The cation is a sufficiently reactive electrophile to undergo an addition-with-elimination reaction with an activated aromatic nucleus.

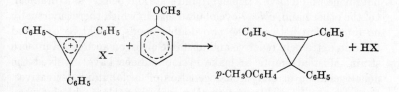

One of the most interesting derivatives of the cyclopropenyl system is cyclopropenone. We can write this compound as an un-saturated ketone or, alternatively, as a dipolar structure.

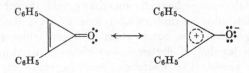

Experimentally, the compound shows very few ketonic properties, and it has a very appreciable dipole moment ($\mu = 5.1$ D; cf. simple ketones, $\mu = 2.4$ D). It also forms very stable salts with mineral acids.

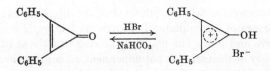

The cyclopropenyl radical, which can be prepared by adding an electron to the cation (experimentally by treating the cation with zinc), shows the typical reactions of a radical and forms a dimer when prepared in solution. Further reactions of this dimer, on heating, are extremely interesting but need not concern us here. Present evidence suggests that the cyclopropenyl anion is most unstable.

Cyclobutadiene, C_4H_4

A prodigious amount of work has gone into attempts to make cyclobutadiene or one of its derivatives. According to the simple molecular-orbital theory we have discussed above, *square*-cyclobutadiene would have a triplet ground state and behave as a diradical. On the other hand, *oblong*-cyclobutadiene, in which there is virtually no interaction between the two double bonds, would be expected to be an extremely reactive and unstable olefin on account of intense strain. Most attempts to make cyclobutadiene have involved the dehalogenation or dehydrohalogenation of cyclobutane derivatives. For example, 3,4-dichlorotetramethylcyclobutene loses chlorine when treated with zinc. No cyclobutadiene can be isolated; but if an acetylene is added an aromatic compound is formed, which indicates the formation of an extremely reactive intermediate, either the square diradical or very strained, oblong, cyclobutadiene.

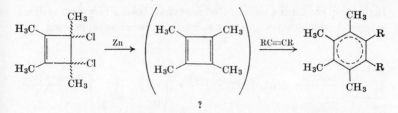

Many attempts have been made to trap the transient cyclobutadiene. The most successful are those in which the cyclobutadiene is complexed to a metal atom. Thus, when the same 3,4-dichlorotetramethylcyclobutene is treated with nickel carbonyl, a crystalline nickel complex of cyclobutadiene is formed. The exact nature of the bonding in this complex and many related compounds will be discussed later in this chapter. When this complex is heated *in vacuo* a mixture of octamethyltricyclooctadiene and octamethylcyclooctatetraene is formed, the former being a precursor of the latter.

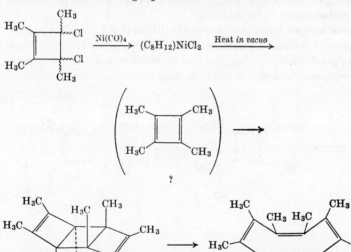

An analogous unsubstituted iron tricarbonyl compound is formed when 3,4-dichlorocyclobutene is treated with iron carbonyl. When this complex is treated with ceric ammonium nitrate at 0° gases are evolved which can be condensed in a trap cooled in liquid nitrogen.

If the trap contains a substituted acetylene then a bicyclohexadiene is formed which rearranges to a substituted benzene on heating.

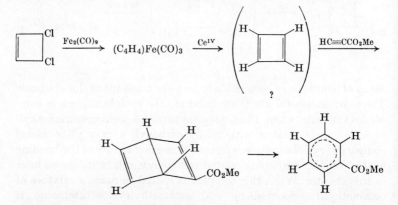

These experiments represent the nearest approach to the preparation of cyclobutadiene so far reported.

A stable cyclobutadiene derivative (diethyl 2,4-bisdiethylamino-cyclobuta-1,3-diene-1,3-dicarboxylate) has been reported but there is very good evidence to suggest that its stability is due to the importance of canonical forms of the type shown.

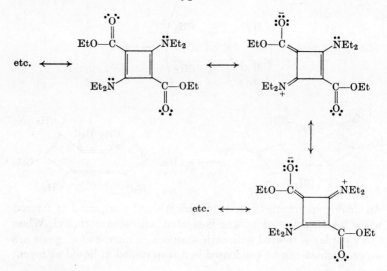

In the hope that a fused benzene ring would stabilize the cyclo-butadiene nucleus, there have been many attempts to make benzo-cyclobutadiene. Dehalogenation and dehydrohalogenation of halogeno-derivatives of benzocyclobutene have yielded a transient intermediate which forms dimers analogous to those formed in the tetramethylcyclobutadiene experiments and which also reacts with dienes to form adducts.

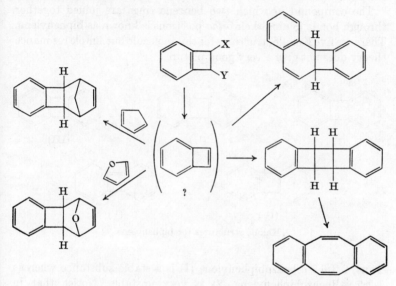

A crystalline cyclobutadiene derivative has been prepared in which the four-membered ring is fused to the 2,3-positions in naphthalene and the remaining corners of the four-membered ring are substituted with phenyl groups.

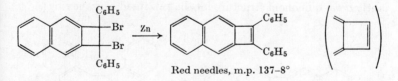

Red needles, m.p. 137–8°

This compound is extremely reactive. It undergoes cycloaddition reactions and when treated with potassium permanganate in acetone yields 2,3-dibenzoylnaphthalene. We saw in Part 2 that there is less double bond character in the 2,3-bond of naphthalene than in the 1,2- or 3,4-bonds. Thus we might regard this compound as being a derivative of cyclobutene with two exocyclic double bonds rather than as a derivative of cyclobutadiene. The positions of the two phenyl groups are also important since the isomeric compound with the phenyl groups attached to the naphthalene nucleus is not stable.

The compound in which two benzene rings are joined together through bonds joining their *ortho*-positions is known as biphenylene. There are five Kekulé structures for this molecule but simple resonance theory does not give a very good picture.

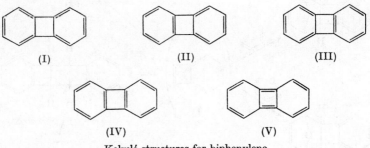

(I)   (II)   (III)

(IV)   (V)

Kekulé structures for biphenylene

Thus, 2,3:6,7-dibenzobiphenylene (**1**) is a stable substance whereas 1,2:7,8-dibenzobiphenylene (**2**) is very unstable. Notice that in formula (**1**) our knowledge of the π electron distribution in naphthalene (cf. Chapter 20 of Part 2) would lead us to expect very little double bond character in the bonds of the four-membered ring. On the other hand, in formula (**2**) we should expect considerable double bond character in a four-membered ring or else a predominantly *ortho*-quinonoid structure for the two fused benzene rings.

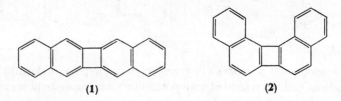

(1)   (2)

Biphenylenes can be prepared from a variety of reactions involving 2,2′-biphenyl derivatives:

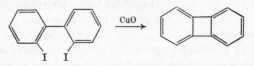

but the yields by this procedure are not good. The other important route to biphenylene proceeds via dehydrobenzene derivatives (see Chapter 22 of Part 2).

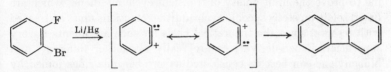

Biphenylene takes part in addition-with-elimination reactions very readily, and substitution occurs in the 2-position. This is not in accord with simple resonance theory if all the canonical forms are given equal weight.

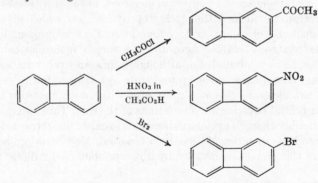

Particularly striking is the coupling of benzenediazonium chloride in the 3-position of 2-hydroxybiphenylene, a reaction very different from the coupling of diazonium salts with 2-naphthol.

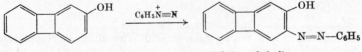

(cf. 2-naphthol)

Simple molecular-orbital theory correctly predicts the site of substitution, and although resonance theory gives the wrong prediction
the observed results can be reconciled with this theory by assuming
that Kekulé structure (I) is a predominant structure.

All the experimental results are consistent with the theoretical
predictions at the beginning of the chapter, namely, that *square*-
cyclobutadiene would be an unstable substance existing as a diradical
while *oblong*-cyclobutadiene would show the reactions of a very highly
strained olefin. It seems probable at the present time that *oblong*-
cyclobutadiene is the state of lower energy, but experiments purport-
ing to prove the multiplicity of transient cyclobutadienes which are
based solely on stereospecific cycloaddition reactions must be treated
with reserve, since the theoretical basis for such arguments is very
superficial. Benzo-derivatives are really altogether different, and
biphenylene can best be considered as two benzene rings joined by
single bonds at their *ortho*-positions. This picture is in accord with the
dimensions of the molecule, the length of the central bonds being
1.52 Å, i.e. about that of a normal single bond.

We saw that the species cyclo-C_3H_3 was necessarily an unstable
radical containing an odd number of electrons, but that by removing
one electron we formed the cyclopropenylium ion which gives a
closed shell of π electrons, and we found that the cyclopropenylium
ion was relatively stable. According to simple molecular-orbital
theory, *square*-cyclobutadiene, although it has an even number of
electrons, should exist in the ground state as a triplet; but if we removed two electrons this would leave us with two π electrons in a
bonding orbital covering the four atoms of the ring. Thus, we might
expect doubly charged cyclobutadiene ions to exist. The tetraphenyl-
cyclobutadiene di-cation (**3**) has been reported. More striking, however, are the properties of 3,4-dihydroxycyclobutene-1,2-dione (**4**).

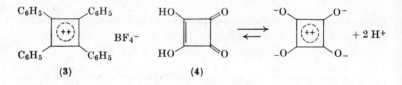

$$\text{(3)} \qquad\qquad\qquad \text{(4)}$$

This substance is a strong dibasic acid. The anion is completely
symmetrical; the four oxygen atoms are equivalent (it should be

added that cyclo-C_5O_5 and cyclo-C_6O_6 ions are also known, indicating that the stability of this ion may be due to other causes as well).

Cyclo-C_5H_5

Cyclopentadiene, C_5H_6, is formed as one of the products in the 'cracking' of light petroleum fractions (see Chapter 17 of Part 1). It behaves as a conjugated diene, undergoing cycloaddition reactions (Diels–Alder reactions), and in fact reacts with itself to form dicyclopentadiene, a reaction in which it differs from both butadiene and cyclohexadiene (butadiene can undergo thermal dimerization to yield vinylcyclohexene).

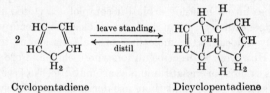

Cyclopentadiene   Dicyclopentadiene

With bases, however, it behaves as a weak acid, forming an anion which will undergo the characteristic reactions of a resonance-stabilized carbanion, e.g. nucleophilic addition to a carbonyl bond or nucleophilic displacement of a halogen.

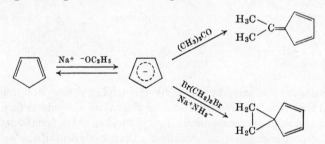

Notice that these reactions are exactly analogous to those of enolate anions described in Chapter 9 of Part 2. The pK of cyclopentadiene is approximately 16, making it more acidic than acetone or even ethanol but appreciably less acidic than ethyl acetoacetate or acetylacetone. At first sight we might be tempted to explain this

acidity in the same way that we explained the acidity of triphenylmethane, namely, that the negative charge be spread over five
carbon atoms. There are several objections to this argument. First
and foremost, cyclopentadiene is a very much stronger acid than
triphenylmethane, and yet we have only five canonical forms and
the charge spread over five atoms compared with forty forms for
the triphenylmethide anion in which the charge is spread over ten
carbon atoms. Further, we have seen that the triphenylcarbonium
ion is also stable, but experiments show that the cyclopentadienyl
cation, $C_5H_5^+$, is extremely unstable. If, instead, we now look at our
diagram for the energy levels of the π orbitals in cyclic compounds,
we can see that when an electron added to the cyclopentadienyl
radical goes into a bonding orbital it completes a closed shell. Thus
the cyclopentadienyl anion has six π electrons occupying three
completely delocalized π molecular orbitals analogous to those of
benzene. However, to describe an ion whose chemical properties are
analogous to those of enolate anions as aromatic does not seem helpful.
We shall see, however, that there are derivatives of the cyclopentadienyl anion which do show aromatic properties.

A number of dipolar compounds have been prepared in which a
cyclopentadienyl anion ring is joined to a cationic centre, forming
a compound called an ylid (Chapter 5, p. 233).

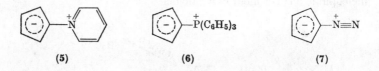

(5)   (6)   (7)

Aromatic character is shown by the triphenylphosphonium ylid
(6) in that it will couple with benzenediazonium chloride to form the
corresponding azo-compound, i.e. it undergoes a characteristic
addition-with-elimination reaction. Diazocyclopentadiene shows
more of the properties of a diazo-compound than of a diazonium
salt and its infrared spectrum shows a strong absorption at 2100
cm^{-1} characteristic of aliphatic diazo-compounds; however, it has
been found to undergo a variety of addition-with-elimination
reactions such as nitration, diazonium coupling, bromination, and
mercuration.

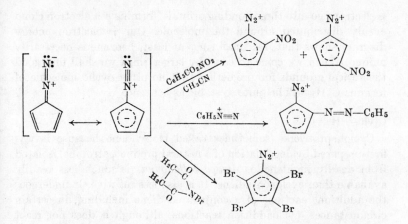

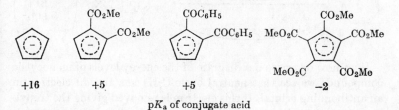

A carbonyl group adjacent to a CH_2 group greatly increases the acidity of the hydrogen atoms on the CH_2 group. It would therefore be reasonable to expect that carbonyl groups attached to cyclopentadiene would increase the acidity of the molecule. This is indeed true; thus dimethyl cyclopentadiene-1,2-dicarboxylate and 1,2-dibenzoylcyclopentadiene both have acidities approaching that of acetic acid while pentamethyl cyclopentadienepentacarboxylate is a strong electrolyte in water (i.e. of mineral acid strength) and reacts with diazomethane to give a *C*-methyl derivative.

pK_a of conjugate acid

Cyclo-C_6H_6

The electronic structure of benzene was discussed in terms of resonance theory in Chapter 13 of Part 1, and in terms of molecular-orbital theory in Chapter 3 of Part 2. There is little we need to add here. The hexagonal shape of the molecule permits the best possible overlapping of the sp^2 hybrid orbitals forming the σ chain. The six

π electrons go into three bonding orbitals, forming a π electron cloud evenly distributed around the molecule. Our π electron energy diagram shows that, for small rings at least, benzene is necessarily unique. When we come to consider larger rings we shall find good theoretical grounds for predicting that no other cyclic molecule of formula C_nH_n will be quite so stable as benzene.

Cyclo-C_7H_7

Cycloheptatriene (sometimes called tropylidene because it was first prepared by degradation of a natural product, atropine, isolated from deadly nightshade, *Atropa belladonna*) is much less readily available than cyclopentadiene. It is also less reactive. It undergoes the addition reactions of a conjugated diene including, in certain circumstances, cycloaddition reactions, although it does not react with itself to form a dimer as does cyclopentadiene. It shows none of the acidic properties of cyclopentadiene but, when treated with triphenylcarbonium fluoroborate or with phosphorus pentachloride, it forms a very stable carbonium ion called the tropylium ion.

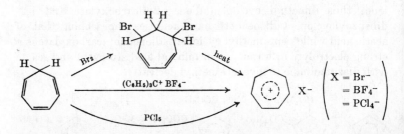

If we now return to our diagram of the energy levels of monocyclic compounds we see that neutral cyclo-C_7H_7 has an odd electron in an antibonding orbital. If this electron is removed giving the tropylium ion, the remaining six π electrons are now all in bonding orbitals so that, just as removal of a proton from cyclopentadiene yields the cyclopentadienyl anion with a closed shell, removal of a hydride anion from cycloheptatriene yields the tropylium ion. Again we have six π electrons in three bonding orbitals distributed evenly over the whole molecule as in benzene and the cyclopentadienyl anion. Just as it would be inappropriate to describe the cyclopentadienyl anion as aromatic, it would be equally unsuitable to describe the tropylium

ion as aromatic. It is a carbonium ion and undergoes the reactions
we should expect of a carbonium ion. It reacts readily with nucleophiles to yield the expected products.

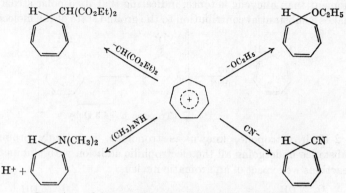

Reactions of the tropylium ion with common nucleophiles

In general, other cations rearrange to the tropylium ion rather than
vice versa, but the tropylium ion, on treatment with neutral permanganate does rearrange, recalling the reactions of aliphatic
carbonium ions.

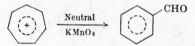

Tropylium ions are weaker electrophiles than the cyclopropenylium
cation but they attack very reactive aromatic nuclei, e.g. phenoxide
ions in an addition-with-elimination reaction.

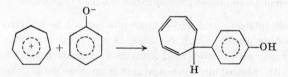

The tropylium ion as the electrophile in an addition-with-elimination reaction

Just as there are derivatives of the cyclopentadienyl anion which
show aromatic properties, so there are derivatives of the tropylium
ion that undergo addition-with-elimination reactions and in general

show aromatic properties. The first compound we should consider
is the ketone cycloheptatrienone otherwise known as tropone. This
compound shows few ketonic properties and has a larger dipole
moment than alicyclic ketones, indicating that its dipolar structure
makes a substantial contribution to the ground state of the molecule.

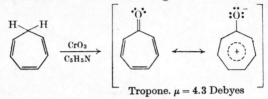

Tropone. $\mu = 4.3$ Debyes

2-Hydroxytropone, known as tropolone, is a truly aromatic
substance undergoing all the electrophilic addition-with-elimination
reactions we expect of an aromatic nucleus.

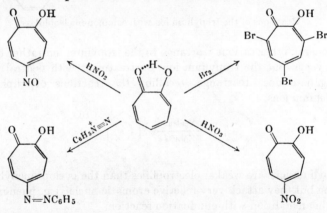

Electrophilic addition-with-elimination reactions of tropolone

The nitro-, nitroso-, and azo-compounds can all be reduced to the
corresponding amine which behaves as a typical aromatic amine.
On treatment with nitrous acid it forms a diazonium salt which will
couple with phenoxide ions and undergo the usual Sandmeyer
reactions. Tropolone is more acidic than phenol and in some of its
reactions could be regarded as the vinylogue of a carboxylic acid
(i.e. although the carbonyl and the hydroxy group are not situated
on the same carbon atom as in a carboxy group they are connected
to each other through a conjugated chain). Thus the hydroxy

group in tropolone, unlike that in phenol, is readily replaced by a chlorine atom on treatment with thionyl chloride. The chlorine atom so introduced is extremely labile and on treatment with an amine the corresponding amino-compound is formed.

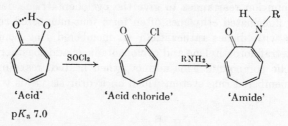

'Acid'   'Acid chloride'   'Amide'

pK_a 7.0

The analogy between these reactions and those of a carboxylic acid, its acid chloride, and its amide will be apparent.

Cyclopentadiene is readily available but cycloheptatriene is not, and therefore a wide variety of reactions have been devised to prepare cycloheptatriene derivatives. The use of carbenes is important, and the photochemical preparation of ethyl cycloheptatriene-carboxylate was described in Chapter 22 of Part 2. Halogenocarbenes, prepared in basic media (see p. 289 of Part 2) can also be used, and a wide variety of such syntheses have been reported.

An interesting reaction is depicted below:

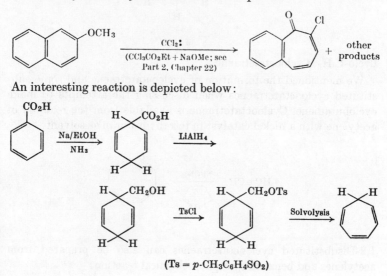

$(Ts = p\text{-}CH_3C_6H_4SO_2)$

Benzoic acid is reduced with sodium in alcohol and liquid ammonia (the Birch reduction), and then further reduced with lithium aluminium hydride to the alcohol. The toluene-*p*-sulphonyl ester of this alcohol is then solvolysed under S_N1 conditions so that the transient carbonium ion rearranges to give the cycloheptatriene derivative. Highly fluorinated ethylenes often form four-membered ring compounds with dienes rather than six-membered ring compounds. Thus, tetrafluoroethylene and cyclopentadiene add together to form a bicyclic compound. This adduct, when heated, rearranges to a seven-membered ring system which on hydrolysis yields tropolone.

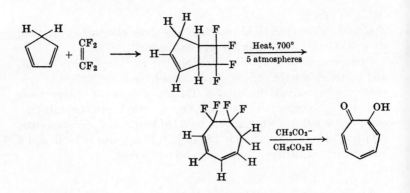

Cyclo-C_8H_8: Cyclooctatetraene

We mentioned the formation of cyclooctatetraene and some substituted cyclooctatetraenes when discussing the attempts to make cyclobutadiene. Cyclooctatetraene is available from the reaction of acetylene with a nickel catalyst in tetrahydrofuran as solvent.

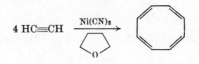

1,2-Disubstituted cyclooctatetraenes can also be prepared from acetylenes and benzene by a photochemical reaction:

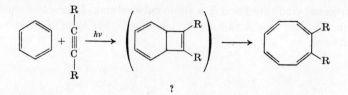

This reaction is interesting because the supposed intermediate also appears to be an intermediate in many of the reactions of cyclooctatetraene. Cyclooctatetraene was first synthesized by Willstätter, starting from a cyclooctane derivative and working through a complete series of addition followed by elimination reactions. This masterly piece of work, carried through before the days of chromatography or spectroscopy, was for many years held in question because the cyclooctatetraene was an unstable polyolefinic substance. We now see that this is exactly what we should expect. According to our very simple theory, planar cyclooctatetraene, in which all the carbon–carbon bonds were of equal length, would exist in a triplet state. There would be six electrons in three bonding orbitals, and the two remaining electrons would go into a degenerate pair of nonbonding orbitals. Simple perturbation theory suggests that complete delocalization of the π electrons would increase and not lower their energy, and experimentally we find that cyclooctatetraene exists as a polyolefin with a puckered shape.

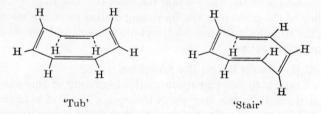

'Tub'   'Stair'

At normal temperatures the 'tub' conformation is believed to predominate in the equilibrium. Since cyclooctatetraene does not show any aromatic properties we need not consider its chemistry in detail here. It is of interest, however, that in its reactions it behaves as though it were in equilibrium with the valence tautomer, postulated as the intermediate in its photochemical preparation from benzene and acetylene. Notice that two of the products actually have a four-membered ring fused to a six-membered ring. In practice this is probably due to a transannular reaction rather than the valence tautomer.

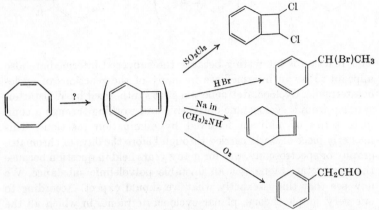

According to our energy level diagram the loss of two electrons from cyclooctatetraene would give the corresponding di-cation, with a closed shell of electrons. Similarly, addition of two further electrons to cyclooctatetraene to give the corresponding di-anion would also give a closed shell. This latter ion has been made and appears to be planar. However, according to the simple molecular-orbital theory that we are using, these two electrons would go into two degenerate non-bonding orbitals. They would therefore do little to increase the stability of the molecule. The four non-bonding electrons would be available to react with any electrophilic species, and the di-anion is extremely reactive.

Cyclo-C_nH_n—Large Rings: The Annulenes

If we return to our discussion at the beginning of this chapter, and particularly to the diagram of the energy levels of π molecular orbitals of the cyclic polyenes, we can see that the molecules with an even number of carbon atoms can be divided into two sets. Molecules with an odd number of carbon atoms must be of radical nature, but in terms of simple resonance theory there is no means of distinguishing between those with an even number of carbon atoms. They apparently should all be equally stable. However, examination of the energy levels predicted by our simple theory shows that this is not so. In the first series, with $n = 2, 6, 10, 14, 18\ldots$ $(n = 4x + 2)$, all the π

electrons go in pairs into bonding orbitals. Neglecting other considerations we should expect these molecules to be stable and to have a chemistry analogous to that of benzene. For the other series of molecules, where $n = 4, 8, 12, 16, 20 \ldots$ ($n = 4x$), our theory predicts that the uppermost π electrons have to occupy a pair of degenerate non-bonding orbitals. Therefore, according to this simple theory the ground state of these molecules should be a triplet or diradical state. Although the molecular-orbital theory on which this argument is based (Hückel theory) is a gross over-simplification it has been extraordinarily successful in predicting the properties of these cyclic polyenes. We have already seen that cyclooctatetraene has none of the properties of benzene. We briefly referred to the slightly more sophisticated theoretical argument which suggests that π electron delocalization, while stabilizing molecules with $4x + 2$ electrons, *destabilizes* molecules with $4x$ π electrons (sometimes called 'anti-aromaticity'). We will now briefly discuss the attempts to make C_{10}, C_{14}, and C_{18} compounds, and the extent to which these compounds are aromatic. These compounds (i.e. ring compounds C_nH_n where $n \geqslant 10$) have been given the name annulenes.

At first sight, cyclo-$C_{10}H_{10}$ (cyclodecapentaene) could exist with a carbon–carbon bond angle of 120°. However, this neglects the hydrogen atoms which would necessarily be inside the ring (see Chapter 13 of Part 1). Experiments suggest that the compound has only transient existence at normal temperatures.

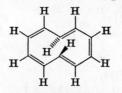

Cyclodecapentaene
(notice the two 'internal' hydrogen atoms)

[12]Annulene would, according to our simple picture, have a triplet ground state, and according to this picture the next compound from which we might expect aromaticity is the [14]annulene. This compound has been prepared. The method of preparation is shown in the scheme. If the compound is to be aromatic it must be planar but, as with the C_{10} compound, the hydrogen atoms in the centre of the ring distort the molecule and it is non-planar.

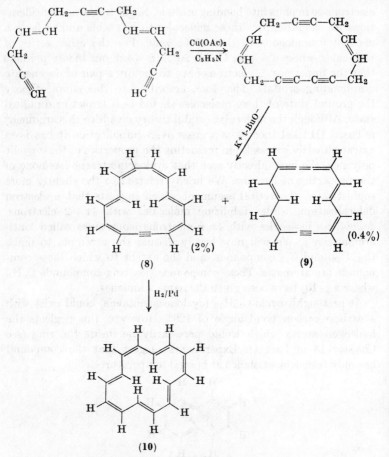

(10)
Brown needles, m.p. 134°
λ_{max} 317 nm (ϵ = 69,000); 376 (25,800)
(ϵ in 10^3 cm^2 mole^{-1} ≡ 10^{-1} m^2 mole^{-1})

It is not the [14]annulene **(10)** that is the interesting compound but
the intermediate **(8)**, which exists in two conformations, and the by-
product **(9)**. **(8)** could be called monodehydro[14]annulene; **(9)**
could be called 1,8-bisdehydro[14]annulene.* In both of these

* These compounds should be called didehydro[14]annulene **(8)** and 1,8-
bisdidehydro[14]annulene **(9)**. Unfortunately it has become current practice
to refer to a 'triple' bond as 'dehydro-' and not 'didehydro-' (cf. Part 2, p. 285).

molecules we have a conjugated cyclic system with 14 π electrons in orbitals of the same symmetry. Notice that the extra π electrons in the acetylenic bond in both molecules and the π electrons in the central allene bond in (**9**) are orthogonal to the remaining π electrons. Both of these molecules can be nitrated with cupric nitrate in acetic anhydride, sulphonated with oleum in dioxan solution, and acylated with acetic anhydride and boron trifluoride to give the corresponding methyl ketone. The [14]annulene (**10**) undergoes none of these reactions. The bisdehydro[14]annulene (**9**) was recovered unchanged on treatment with an excess of maleic anhydride in boiling benzene. On similar treatment the greater part of the monodehydro[14]annulene was recovered unchanged whereas almost all the [14]annulene (**10**) had reacted. The bisdehydro[14]annulene (**9**) is a particularly interesting compound. It has been shown to be planar and it is extremely stable. Notice that we can draw Kekulé structures or alternatively represent the completely delocalized π orbitals as shown.

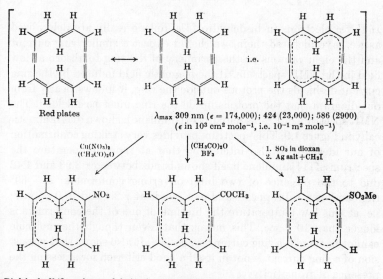

Bisdehydro[14]annulene and derivatives prepared by electrophilic addition-with-elimination reactions.

(nitration, Friedel–Crafts acylation, and sulphonation)

Spectroscopic properties of these compounds are interesting and important. First, notice the greatly increased intensity of the ultraviolet absorption band of the bisdehydro[14]annulene compared with that of [14]annulene itself. In Chapter 25 of Part 2 we attributed the fact that in the NMR spectrum of benzene the protons absorb at low field to the presence of a ring current induced by the applied magnetic field on the closed loop of the π electrons (Figure 1.8).

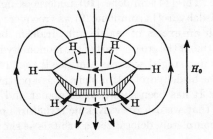

Figure 1.8.   Deshielding of aromatic protons by the ring current effect.

If the π electrons in bisdehydro[14]annulene really are delocalized as we have pictured them, we should expect a similar ring current and the eight protons on the periphery of the ring to absorb at low field in the NMR spectrum. If the magnetic field induced by the ring current deshields the protons outside the ring, it follows clearly from our diagram that the protons inside the ring must be shielded. The NMR spectrum of 1,8-bisdehydro[14]annulene is shown diagrammatically in Figure 1.9. This spectrum provides very striking confirmation of our ideas. It is also interesting that at room temperature the spectrum of [14]annulene itself shows bands between τ 3.93 and 4.42 due to the presence of two interconverting conformers; at $-60°$ the band at τ 4 disappears and new bands at τ 2.4 and 10.0 appear; i.e. at this low temperature the lifetime of one of the conformers is longer than 10^{-8} sec. This means that, even though the molecule cannot be planar, a ring current is possible. It also shows that possession of a ring current is not in itself a good criterion for assessing the presence of aromaticity.

[16]Annulene would not, according to our picture, be an aromatic compound but a cyclic polyolefin like cyclooctatetraene. This com-

pound has been prepared; the molecule is non-planar and in the solid state shows almost complete bond alternation. Its chemical properties are those of an unsaturated cyclic olefin. More interesting is the [18]annulene which has been synthesized as shown on p. 32.

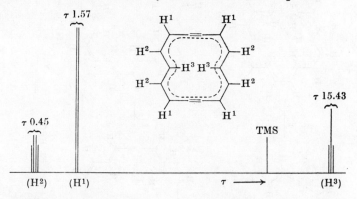

Figure 1.9. NMR (60 MHz) spectrum of 1,8-bisdehydro[14]annulene

We have drawn [18]annulene as an aromatic compound (denoted by the broken ring) but its aromaticity is far less marked than that of bisdidehydro[14]annulene. First attempts to carry out electrophilic addition-with-elimination reactions were unsuccessful, and although nitration with cupric nitrate and acetic anhydride has now been reported, the molecule is appreciably less stable than the bisdidehydro[14]annulene. In the crystalline state it is very nearly planar and its NMR spectrum shows evidence of the presence of a ring current having absorptions at low fields (τ ca. 1.1) due to the outside protons and at high fields (τ ca. 11.8) due to the internal protons. Although [18]annulene shows only limited aromatic properties, these are in contrast to the fact that [24]annulene is a very unstable compound, decomposing in the air and having no aromatic properties of any kind. Its NMR spectrum at room temperature shows only one broad band at τ 3.16 due to all the protons.

The lack of real aromaticity of the [18]annulene may be partly due to the non-bonded repulsions between the hydrogen atoms in the centre of the ring. It may also be due to the inadequacy of the simple molecular-orbital theory on which we have been basing our discussion. More sophisticated theories suggest that as the rings get larger

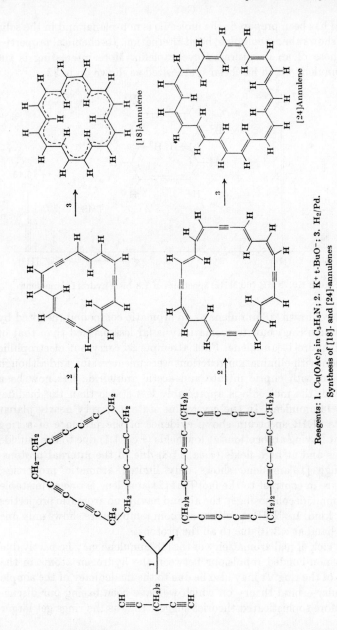

Reagents: 1. $Cu(OAc)_2$ in C_5H_5N; 2. K^+ t-BuO$^-$; 3. H_2/Pd.

Synthesis of [18]- and [24]-annulenes

the tendency for the bonds to alternate into double and single bonds becomes greater.

We will now consider bridged aromatic systems. Supposing a molecule was constructed in which the periphery of the molecule was made up of sp^2 hybridized carbon atoms forming a closed cyclic π orbital system, would it be possible to bridge across this ring with a saturated chain? Several examples of such molecules have been prepared. A particularly interesting example is 1,6-methylenecyclodecapentaene. We have seen that [10]annulene (cyclodecapentaene) exists only transiently, but this bridged derivative has been prepared and it does show aromatic properties.

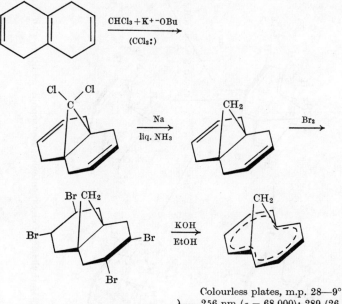

Colourless plates, m.p. 28—9°
λ_{max} 256 nm ($\epsilon = 68,000$); 289 (26,200)
(ϵ in 10^3 cm^2 mole$^{-1} \equiv 10^{-1}$ m^2 mole^{-1})

The compound has been brominated, nitrated, and acylated.

The NMR spectrum confirms the presence of a ring current, the peripheral protons absorbing at low field (τ 2.5—3.2) and the methylene bridge, which is above the centre of the ring, at high field (τ 10.5). The bromo-compound has been converted into the corresponding

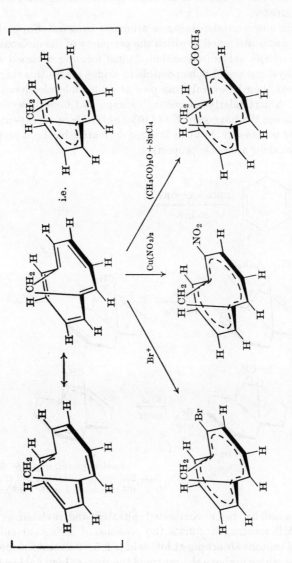

1,6-Methylenecyclodecapentaene; electrophilic addition-with-elimination reactions

Grignard reagent and so into the carboxylic acid. This acid has been converted through the acid chloride and azide by a Curtius rearrangement into the isocyanate and, finally, into the amino-compound. The amine will not undergo diazotization but the hydrochloride on hydrolysis yields the hydroxy-compound, which is particularly interesting because it is in equilibrium with the keto-form in which the 1,6-bond is present.

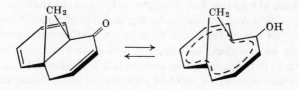

A closely analogous molecule is *trans*-15,16-dimethyldihydropyrene. The periphery of the molecule is identical with what we should expect for [14]annulene if the latter were planar. In the very centre of this conjugated ring we have two bridging carbon atoms each with a methyl group attached to it.

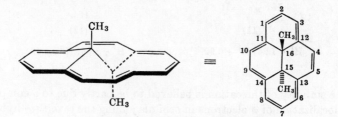

trans-15,16-Dimethyldihydropyrene
Green needles, m.p. 119°; λ_{max} 377 nm (ϵ = 87,000); 463 (6000); 641 (330)
(ϵ in 10^3 cm^2 mole^{-1}, i.e. 10^{-1} m^2 mole^{-1})

trans - 15,16 - Dimethyldihydropyrene undergoes electrophilic addition-with-elimination reactions, behaving as a very reactive hydrocarbon. With N-bromosuccinimide it forms the 2,7-dibromoderivative. It forms the 2-nitro-compound on treatment with cupric nitrate in acetic anhydride, and the 2-benzoyl compound on treatment with benzoyl chloride and aluminium chloride. The NMR spectrum of this compound is very striking. The hydrogen atoms on the two methyl groups absorb at very high field (a sharp singlet at τ 14.25) while the hydrogen atoms on the periphery absorb at low field (τ 1.33—2.02).

Our discussion of these bridged compounds leads us on to consider briefly a phenomenon given the most unsuitable name 'homo-aromaticity'. The term 'homo' has had a somewhat chequered history in organic chemistry, but it is generally used to indicate a chain extension or a ring expansion by the addition of a methylene (CH_2) group. The use of the same term in connection with aromaticity can best be explained by taking a particular example. Cycloocta-trienone (**11**) forms very stable salts (**12**) with strong acids. It will be seen that this cation could be regarded as being analogous to the cation formed when tropone is treated with acid, except that there is a CH_2 bridge across the 1,7 carbon atoms.

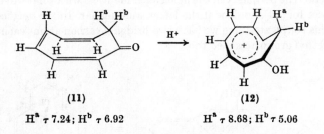

(**11**)   (**12**)

H^a τ 7.24; H^b τ 6.92 H^a τ 8.68; H^b τ 5.06

The stability of the cation is believed to be partly due to a complete delocalization of π electrons in orbitals joining the seven sp^2 hybridized carbon atoms which are believed to form a planar seven-membered ring. The NMR spectrum of the cation supports this view. The methylene protons of the cyclooctatrienone (**11**) have the absorptions shown. In the cation the inside proton (H^a) and the outside proton (H^b) absorptions have shifted. The most likely explanation of this shift is the presence of a ring current. A number of metal complexes have been prepared whose behaviour can best be interpreted by assuming the presence of π molecular orbital bonding between two carbon atoms that are joined by a methylene bridge but have no σ bond in addition to their π bond.

Polycyclic Compounds*

Apart from naphthalene we can devise on paper a large number of bicyclic compounds containing alternate single and double bonds. As with benzene, we can usually draw at least two canonical forms for these compounds, and we might, in terms of simple resonance theory, expect that these compounds would show aromatic properties. We shall consider only three of them, pentalene, azulene, and heptalene.

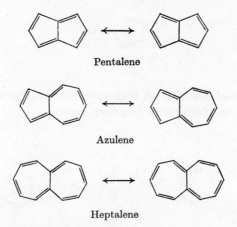

Pentalene

Azulene

Heptalene

The energy levels of the molecular orbitals for these molecules can be calculated by Hückel theory but they do not follow a simple pattern as do the monocyclic compounds. Thus, whereas the possession of

---

*According to the original electronic theories of Lapworth, any molecule that can be drawn with a series of alternating single and double bonds forming a completely conjugated system such as benzene (A), pentalene (B), or fulvene (C) may suitably be called an 'alternate' hydrocarbon.

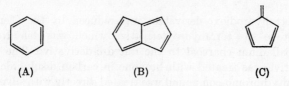

(A)   (B)   (C)

Unfortunately, this very convenient nomenclature clashes with the term 'alternant' which is derived from quantum-mechanical ideas and which excludes any ring system with an odd number of atoms.

$(4x + 2) \pi$ electrons in a monocyclic system is usually associated with, at least, quasi-aromatic character, no such condition applies to polycyclic compounds.

We shall begin by considering two fused five-membered rings in the form of pentalene. It is perhaps worth noting, before discussing this compound further, that the alternate hydrocarbon fulvene (N.B. not 'alternant') is an extremely reactive conjugated diene.

Fulvene

Pentalene itself has not yet been prepared although derivatives of the dibenzo-derivative have long been known.

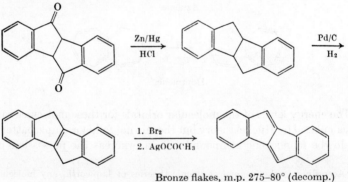

Bronze flakes, m.p. 275–80° (decomp.)
λ_{max} 281 nm ($\epsilon = 69{,}000$); 394 (12,000); 410 (15,000)
(ϵ in 10^3 cm^2 mole^{-1}, i.e. 10^{-1} m^2 mole^{-1})

The tetrahydrodioxo-derivative was reduced by the Clemmensen reagent to the tetrahydro-derivative which was dehydrogenated over palladium–charcoal to the dihydro-derivative. The dihydro-derivative was treated with bromine in carbon disulphide and the resultant dibromo-compound was treated directly with silver acetate to yield dibenzo[*a,e*]pentalene. Dibenzopentalene behaves as a conjugated diene; it reacts very rapidly with ozone, is easily hydrogenated, and reacts with bromine to give an addition compound.

Hexaphenylpentalene has been synthesized from 1,2,3-triphenyl-cyclopentadiene, by condensation with 1,2,3-triphenylpropenone. The resultant dihydrohexaphenylpentalene was readily dehydrogenated by treatment with *N*-bromosuccinimide.

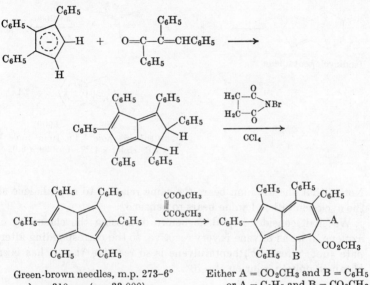

Green-brown needles, m.p. 273–6°
λ_{max} 310 nm (ϵ = 33,000);
380 (9800); 720 (900)
(ϵ in 10^3 cm^2 mole^{-1}, i.e. 10^{-1} m^2 mole^{-1})

Either A = CO_2CH_3 and B = C_6H_5
or A = C_6H_5 and B = CO_2CH_3

In the crystalline form hexaphenylpentalene is quite stable but solutions are sensitive to the air. Its lack of aromatic character is shown by its reaction with dimethyl acetylenedicarboxylate; the adduct undergoes spontaneous ring-expansion to yield an azulene derivative.

Although pentalene itself is unknown, the di-anion has been prepared. Isodicyclopentadiene (prepared in two steps from dicyclopentadiene), when heated in a nitrogen stream at 570°, breaks down into ethylene and dihydropentalene, together with small amounts of other compounds. Dihydropentalene, treated with n-butyllithium, yields the dilithium salt of the pentalenylene di-anion. The salt is insoluble in pentane, but it is moderately soluble in tetrahydrofuran in which solution its NMR spectrum has been examined. Only two bands are observed, a triplet at τ 4.27 and a doublet at 5.02 with identical splittings of 3.0 Hz.

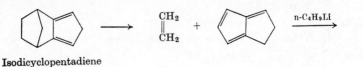

Isodicyclopentadiene

λ_{max} 296 nm (ϵ = 6000)
(ϵ in 10^3 cm^2 mole^{-1}, i.e. 10 m^2 mole^{-1})
NMR triplet τ 4.27; doublet τ 5.02

Notice that the di-anion bears the same relation to naphthalene as the cyclopentadienyl anion bears to benzene.

We shall consider heptalene before discussing the chemistry of azulene. Just as fulvene is very reactive, so the corresponding alternate (not 'alternant') heptafulvene is so reactive that it has been obtained only in solution at $-70°$.

Heptafulvene

The cycloheptatriene synthesis described on p. 24 was used for the synthesis of heptalene. Naphthalene-1,5-dicarboxylic acid is reduced by a Birch reduction. The dihydronaphthalene derivative is further reduced with lithium aluminium hydride to a diol which is treated with toluene-p-sulphonyl chloride to yield a di-toluene-p-sulphonyl ester. The ester undergoes unimolecular solvolysis with molecular rearrangement to yield a mixture of dihydroheptalenes. These are dehydrogenated with triphenylmethyl fluoroborate to yield bright **yellow** crystalline heptalenium fluoroborate. On treatment with triethylamine this salt is converted into heptalene, a very **unstable** liquid which polymerizes when warmed or when treated with oxygen.

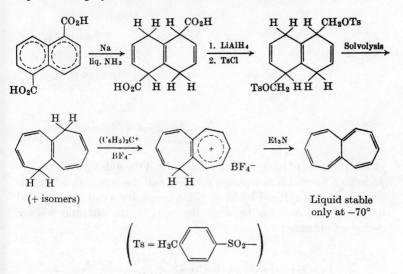

$$\left(Ts = H_3C\!\!-\!\!\bigcirc\!\!-SO_2- \right)$$

By analogy with the pentalene di-anion we might expect that the heptalene di-cation would also be stable. At the present time this compound has not been prepared.

Thus, neither the alternate hydrocarbon with two five-membered rings fused together nor the one with two seven-membered rings fused together shows any aromatic properties, in spite of the existence of two Kekulé structures. Naphthalene, with two six-membered rings fused together, has already been discussed in Part 2, and this leaves us with azulene, the alternate formed by fusing a five-membered to a seven-membered ring.

Azulene derivatives were first discovered in the residues from the distillation of oils extracted from plants. It is now known that these are derived from natural products called sesquiterpenes (see Chapter 5 of Part 4) as the result of dehydrogenation. Azulene itself is most readily prepared from cyclopentadiene and a derivative of glutaconic aldehyde. The glutaconic aldehyde is readily prepared from the heterocyclic base pyridine by treating its dinitrophenyl quaternary salt with methylaniline.

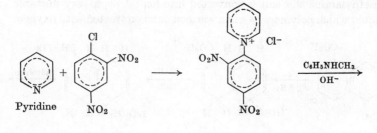

$$O=CH-CH=CH-CH=CH-N\big<{}^{CH_3}_{C_6H_5}$$

In the presence of base, cyclopentadiene and the aldehyde derivative undergo a normal Knoevenagel reaction and the resulting condensation product is cyclized by heat. (This preparation can be simplified by the direct reaction between the pyridinium salt and sodium cyclopentadienide.)

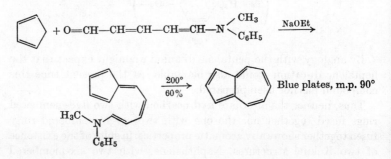

A wide variety of other syntheses of azulene and azulene derivatives has been devised. We shall not discuss these syntheses here but simply note that azulenes have been prepared with alkyl groups in each of the possible positions.

The striking feature of azulene is its intense blue colour. It is a stable compound and, apart from its visible spectrum, its other striking physical property is a dipole moment of 1 Debye. We can account for this dipole moment if we assume that there is some tendency for the five-membered ring to gain an electron and the seven-membered ring to lose an electron, to give a cyclopentadienyl anion fused to a tropylium cation.

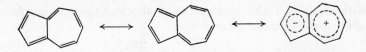

There is, of course, no complete transfer of charge in the ground state as depicted in the right-hand formula. Azulene behaves as a base: it dissolves in 60% sulphuric acid to give a yellow solution, and on dilution the blue hydrocarbon is reprecipitated.

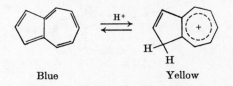

Blue   Yellow

Azulene undergoes electrophilic addition-with-elimination reactions extremely readily.

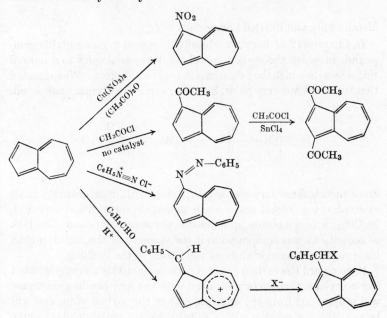

Azulene; electrophilic addition-with-elimination reactions

With really powerful nucleophiles it also undergoes nucleophilic addition-with-elimination reactions.

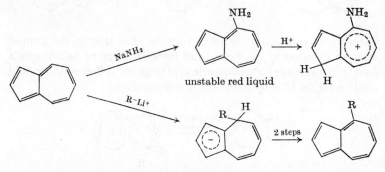

unstable red liquid

Notice that, in conformity with the dipolar character of the molecule, electrophiles attack the five-membered ring and nucleophiles attack the seven-membered ring.

Metallocenes and Related Compounds

In Chapter 12 of Part 1 we briefly discussed organometallic compounds in which the carbon–metal bond was analogous to a normal single bond in which the two atoms shared two electrons. We suggested that this bond was very polar, but otherwise it is a normal single bond.

$$\left[M-C\begin{smallmatrix} R \\ R \\ R \end{smallmatrix} \longleftrightarrow M^+ \ {}^-:C\begin{smallmatrix} R \\ R \\ R \end{smallmatrix} \right]$$

Some metals form very stable compounds with unsaturated organic molecules; e.g. nickel and carbon monoxide form nickel carbonyl, $Ni(CO)_4$, a very volatile, quite stable covalent compound. The lack of polarity in this compound and the strength of the metal–carbon bond require some new ideas on the nature of the bonding.

If we regard the carbon atom of carbon monoxide as sp hybridized like acetylene, then an orbital, filled with two non-bonding electrons, will project out from the opposite side of the carbon atom and will be available for overlap with a suitable vacant metal orbital to give a polar σ bond. Now the metal may have filled d orbitals (see Chapter

Vacant
metal orbital

Filled
carbon sp^1 orbital

σ-bond
(C-atom donating)

5) which can interact with the vacant antibonding π orbitals of the carbon monoxide.

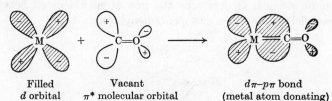

Filled
d orbital

Vacant
π^* molecular orbital

$d\pi-p\pi$ bond
(metal atom donating)

The overall effect is that of a non-polar double bond.

If the orbitals of carbon monoxide can interact with atomic orbitals of the metal atom to give a bonding interaction, we may expect a somewhat similar interaction between olefins and metals; e.g. silver–olefin complexes have been depicted as formed through the

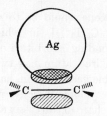
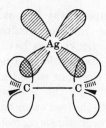

Vacant $5s$ orbital
of silver overlapping
with filled π orbital
of an olefin

Filled $4d$ orbital
of silver overlapping
with vacant π^* orbital
of an olefin

combined bonding action of the vacant $5s$ orbital of the silver overlapping the filled π orbital of the olefin together with the bonding action of a filled $4d$ orbital† of the silver overlapping the vacant π^* orbital of the olefin. In this chapter we are not concerned with

† The nature and shapes of d orbitals are discussed in Chapter 5.

the nature of the bonding in organometallic compounds but with the chemical reactions of molecules where reactivity parallels that of benzene. The metallocenes are 'complexes' formed between cyclic unsaturated organic residues and a transition metal, many of which show typical 'aromatic' properties.

Metallocenes are typified by ferrocene (dicyclopentadienyliron), the first compound of this type to be fully characterized and still the most studied. In ferrocene the iron atom is located between the parallel planes of two cyclopentadienyl rings.

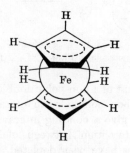

The simplest way of looking at ferrocene would be to regard it as ferrous di(cyclopentadienide), $[Fe^{2+}(C_5H_5^-)_2]$, in which both cyclopentadienyl rings have all their bonding π orbitals occupied. However, we should also note that an iron atom has twenty-six planetary electrons while the nearest noble gas, krypton, has thirty-six. Thus, if all the π electrons of the cyclopentadienyl rings are shared with the iron atom, the iron has a share in thirty-six electrons. If this latter picture is correct we might expect an analogous sandwich compound involving a chromium atom, with twenty-four planetary electrons, and two benzene rings.

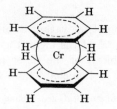

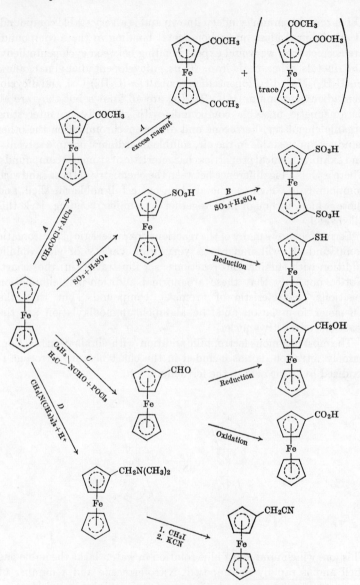

Electrophilic addition-with-elimination reactions undergone by ferrocene.
A, Friedel–Crafts acylation; *B*, sulphonation; *C*, Vilsmeier formylation;
D, Mannich aminoalkylation

Dibenzenechromium is indeed known and is a very stable compound.
If our original ideas on carbon–metal bonding in these compounds
are correct then we would expect bonding between cyclopentadienyl
and metals other than iron. Thus, dicyclopentadienylmanganese
$[\pi\text{-}(C_5H_5)_2Mn]$, dicyclopentadienylcobalt $[\pi\text{-}(C_5H_5)_2Co]$, and dicyclo-
pentadienylchromium $[\pi\text{-}(C_5H_5)_2Cr]$ are all known but they are all
paramagnetic unstable compounds with virtually no interesting
organic chemistry. Ferrocene and dibenzenechromium on the other
hand are quite stable in the air, soluble in ordinary organic solvents,
and exhibit chemical properties characteristic of aromatic compounds.
There is thus a big difference between the chemistry of those sandwich
compounds in which the metal atom has a full noble-gas shell, and
those which, although held together by similar bonding, lack this
filled shell.

Ferrocene shows many of the reactions characteristic of an aromatic
compound. It will undergo a very wide variety of electrophilic
addition-with-elimination reactions as illustrated in the chart.
Notice not only that these are normal addition-with-elimination
reactions characteristic of aromatic compounds, but that the
Vilsmeier formylation and the Mannich aminoalkylation are the
reactions of reactive nuclei.

The most common electrophilic addition-with-elimination reaction,
namely nitration, is not included in the chart because ferrocene is
oxidized by nitric acid to the ferricinium ion.

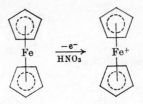

This ion, which produces a blue solution in water, lacks the noble-gas
shell and is rapidly decomposed. Nitroferrocene and a number of
ferrocene derivatives can be prepared from the lithium derivative
as depicted in the second chart.

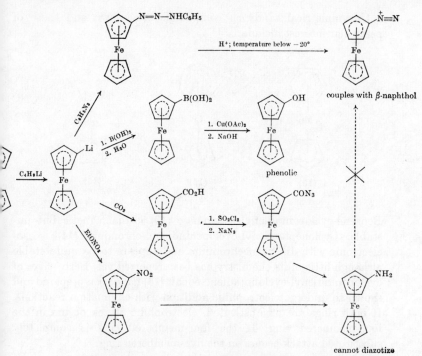

Characteristic 'aromatic' reactions of ferrocenyl-lithium

Ruthenocene [π-$(C_5H_5)_2$Ru] and osmocene [π-$(C_5H_5)_2$Os] show similar chemical properties to ferrocene. They have stronger ring-to-metal bonding and need more vigorous conditions for electrophilic addition-with-elimination reactions.

Dibenzenechromium will not undergo electrophilic attack, but substitution of dibenzenechromium is possible via metallation.

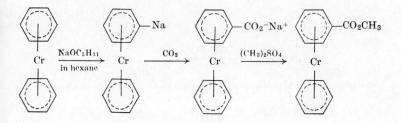

Unsymmetrical sandwich compounds are known and those of particular interest include:

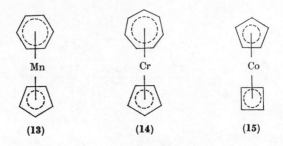

(13)   **(14)**   **(15)**

Benzene-cyclopentadienyl-manganese **(13)** is diamagnetic but unstable. Cycloheptatrienyl-cyclopentadienyl-chromium **(14)** is isoelectronic with dibenzenechromium and appears to be quite stable although little of its chemistry has been reported yet. Derivatives of cyclopentadienyl-cyclobutadiene-cobalt **(15)** have been prepared and shown to undergo electrophilic addition-with-elimination reactions. If both rings are unsubstituted, electrophilic attack occurs in the four-membered ring. If the four-membered ring is completely substituted attack occurs on the five-membered ring.

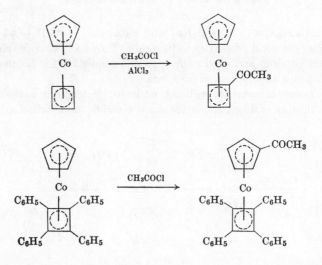

Closely related to the metallocenes, and also showing considerable 'aromatic character' in their chemical reactions, are the 'piano stool' compounds.

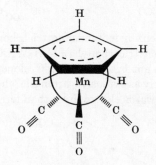

Cyclopentadienylmanganese tricarbonyl; a 'piano stool' compound

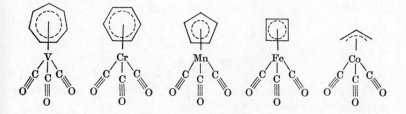

Cycloheptatrienylvanadium tricarbonyl is a dark green diamagnetic substance but not very stable. Benzenechromium tricarbonyl is a very stable, yellow diamagnetic substance, which undergoes electrophilic addition-with-elimination reactions such as Friedel–Crafts acylation only with difficulty. This resistance to attack by electrophiles is due to the electron-withdrawing influence of the metal. Thus the carboxylic acid $[\pi\text{-}(C_6H_5CO_2H)Cr(CO)_3]$ is a stronger acid than p-nitrobenzoic acid, and the amine $[\pi\text{-}(C_6H_5NH_2)Cr(CO)_3]$ is a very weak base indeed. As might be expected, the reluctance to undergo electrophilic attack is matched by a readiness to undergo nucleophilic addition-with-elimination reactions.

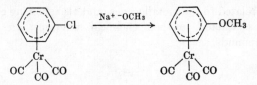

The corresponding benzenemolybdenum tricarbonyl [π-C$_6$H$_6$Mo(CO)$_3$] and benzenetungsten tricarbonyl [π-C$_6$H$_6$W(CO)$_3$] are known but their chemistry has not been explored. Cyclopentadienylmanganese tricarbonyl has nearly as extensive 'aromatic' chemistry as ferrocene. It is less reactive towards electrophiles than ferrocene.

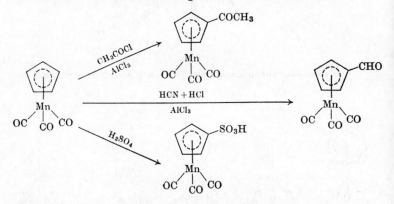

Cyclobutadieneiron tricarbonyl [π-C$_4$H$_4$Fe(CO)$_3$], on the other hand, is more reactive than ferrocene.

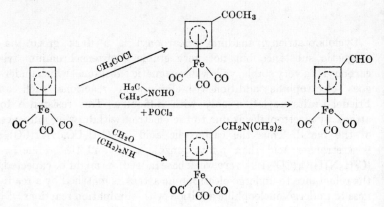

The formation of transient cyclobutadiene by treating this latter compound with ceric salts has been described on page 12. As yet, no electrophilic addition-with-elimination reactions of π-allylcobalt tricarbonyl have been reported although the Friedel–Crafts acylation of π-butadiene-metal carbonyl complexes has been achieved.

In the 'piano stool' compounds we have discussed, the metal obtains a noble-gas shell through the donation of the π electrons from the arene and six electrons from the three carbonyl groups. We can obtain another similar series of compounds in which we keep the arene nucleus constant and vary the number and nature of the inorganic ligands (NO gives 3 electrons; CO gives 2; I gives 1).

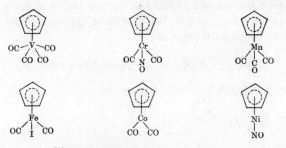

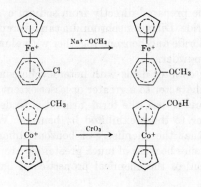

Diamagnetic cyclopentadienyl–metal complexes

The first three of these compounds undergo electrophilic addition-with-elimination reactions, in particular Friedel–Crafts acylation.

Aromatic reactions are also exhibited by ionic 'sandwich' compounds in which the metal atom has a share in sufficient electrons to provide a noble-gas shell.

There is thus a very extensive range of organometallic compounds showing the chemical properties characteristic of benzene. The present discussion has been concerned only with this aspect of their chemistry; full acounts of their 'inorganic chemistry' will be found elsewhere.

Preparative details are outside the scope of these books. We may simply note that most of these π-arene-metal complexes are formed by ligand exchange:

$$Ar(2n\ \pi\ \text{electrons}) + M(CO)_m \rightarrow \pi\text{-ArM(CO)}_{m-n} + nCO$$

e.g. $$C_6H_6 + Cr(CO)_6 \rightarrow \pi\text{-}C_6H_6Cr(CO)_3 + 3CO$$

If the arene has an odd number of carbon atoms, fairly high temperatures may be necessary to eliminate hydrogen from the corresponding conjugated polyene:

$$2Fe(CO)_5 + 2C_5H_6 \xrightarrow{250°} \pi\text{-}(C_5H_5)_2Fe + 5CO + H_2$$

In some cases hydride transfer is possible:

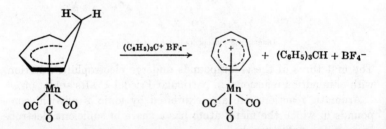

Ferrocene can be prepared directly from sodium cyclopentadienide and ferrous chloride. Dibenzenechromium can be prepared by heating chromium trichloride and benzene together with aluminium chloride and aluminium powder.

The following chapter deals with heterocyclic compounds, which show aromatic character to a greater or lesser extent. We have used, and shall continue to use, the term 'aromatic' to describe chemical properties similar to those exhibited by benzene. We have seen in Parts 1 and 2 that the chemistry of benzene differs from that of conjugated polyenes because of much greater electron delocalization. **This 'explanation' of the chemical properties of benzene has been** taken over by some authors and used as a definition of aromaticity. As we see it, the term 'aromatic' is of value because it defines a type of chemical reactivity. To use either a theoretical concept or a specific physical property which is not associated with *any* kind of chemical reactivity as a criterion of aromaticity seems to us an unhelpful exercise. However, readers will come across such definitions.

Bibliography

D. M. G. Lloyd, *Carbocyclic Non-benzenoid Aromatic Compounds*, Elsevier, Amsterdam, 1966.

M. P. Cava and M. J. Mitchell, *Cyclobutadiene and Related Compounds*, Academic Press, London and New York, 1967.

D. Ginsburg (Ed.), *Non-benzenoid Aromatic Compounds*, Interscience, New York, 1959.

G. M. Badger, *Aromatic Character and Aromaticity*, Cambridge University Press, 1969.

P. L. Pauson, *Organometallic Chemistry*, Edward Arnold, London, 1967.

M. Rosenblum, *Chemistry of the Iron Group Metallocenes*, Interscience, New York, 1965.

Problems

1. Suggest a synthesis of $C_9H_9^-$, and discuss the properties you would expect for such an ion.

2. Suggest a synthesis of

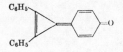

and comment on possible 'aromaticity'.

3. Starting from 2-hydroxybiphenylene suggest a synthesis of the corresponding quinone

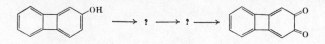

4. How does azulene react with: (a) 70% H_2SO_4; (b) $(CF_3CO)_2O$; (c) $LiNH_2$; (d) $CH_3CHO + H^+$.

5. Suggest possible methods for the synthesis of

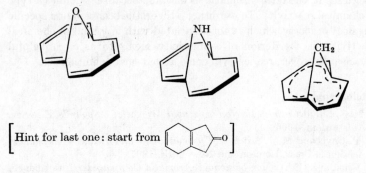

$$\left[\text{Hint for last one: start from} \quad \text{O} \right]$$

CHAPTER 2

Heterocyclic Aromatic Compounds

Pyridine (6-membered ring containing one nitrogen atom)

In terms of our pictorial molecular-orbital theory we have considered benzene as made from six carbon atoms arranged hexagonally. These atoms were considered to be joined to each other by σ bonds formed from overlapping sp^2 hybrid atomic orbitals and each carbon atom bonded to a hydrogen atom by a σ bond formed by the overlap of a carbon sp^2 hybrid atomic orbital and a hydrogen $1s$ atomic orbital. In addition, we visualized the unhybridized p_z atomic orbital of each carbon atom overlapping with the corresponding orbitals on the two adjacent carbon atoms. This overlap led to six π orbitals, three bonding and three antibonding. Each carbon atom in the ring contributed one electron to the π bond system so that each of the three bonding orbitals was filled. Let us now consider what would happen if we replaced one of the carbon atoms together with its adjacent hydrogen atom by a nitrogen atom. Since it is possible to

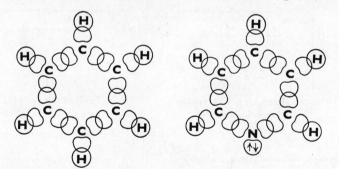

Overlapping sp^2 and $1s$ atomic orbitals of carbon, hydrogen, and nitrogen leading to σ bonds in benzene and pyridine.

visualize a nitrogen atom sp^2 hybridized, the σ bonds of the ring can remain virtually unaltered. Similarly, the unhybridized p_z atomic orbital on the nitrogen will overlap with the corresponding p atomic orbitals of the two adjacent carbon atoms so that again there must be six π orbitals formed by the interaction of the p_z atomic orbitals of the five carbon atoms and of the one nitrogen atom.

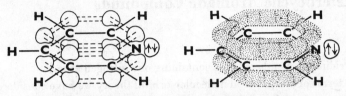

π-Electron 'clouds' in pyridine (cf. Part 2, p. 34).

We should expect this compound, called pyridine, to be aromatic and to show chemical properties similar to those of benzene. In benzene four of the π orbitals occur as degenerate pairs ψ_2 and ψ_3, forming a bonding degenerate pair, and ψ_4 and ψ_5, forming an anti-bonding degenerate pair. This degeneracy is lost in pyridine owing to the influence of the heteroatom (i.e. $\psi_1 < \psi_2 < \psi_3 < \psi_4 < \psi_5 < \psi_6$).

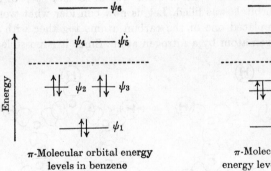

π-Molecular orbital energy π-Molecular orbital
levels in benzene   energy levels in pyridine

Hückel orbitals

This slight alteration in the π orbital energy levels does not affect the fact that there is a π electron cloud above and below the plane of the ring. The important difference from benzene is the sp^2 hybrid orbital of the nitrogen atom projecting out of the ring which is fully occupied by two electrons and does not form a bond with a hydrogen atom. The lone pair is analogous to the lone pair of ammonia and is capable of forming a bond with a proton; i.e. pyridine is a base. It is a weaker base than ammonia or triethylamine, presumably because the nitrogen lone pair is in an sp^2 orbital (i.e. an orbital with more s character and closer to the nucleus) compared with an sp^3 orbital in the other compounds. Pyridine is a strong-smelling colourless liquid (b.p. 114°) completely miscible with water.

Table 2.1. Dissociation constants of the conjugate acids of pyridine, ammonia, and triethylamine

Conjugate acid	Base	pK_{BH^+}
	$+ H^+$	5.2
H_4N^+ $\rightleftharpoons$	$H_3N + H^+$	9.3
$(C_2H_5)_3\overset{+}{N}H$ $\rightleftharpoons$	$(C_2H_5)_3N + H^+$	10.8

Realizing that pyridine is simultaneously an aromatic compound and a tertiary base, let us now consider its chemical properties. The first striking property about benzene is its great stability and low reactivity compared with what we might have predicted for the hypothetical cyclohexatriene (cf. Part 1, p. 127). In particular, benzene is resistant to oxidation by reagents such as permanganate or chromic acid and it does not readily undergo addition reactions such as hydrogenation. As a first approximation it can be said that pyridine undergoes addition reactions more readily than benzene; on the other hand pyridine is more resistant to oxidizing agents than benzene. Thus, 2,3-benzopyridine (called quinoline) can be reduced to 1,2,3,4-tetrahydroquinoline and oxidized to pyridine-2,3-dicarboxylic acid (quinolinic acid).

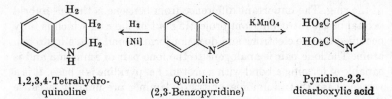

1,2,3,4-Tetrahydro-   Quinoline   Pyridine-2,3-
quinoline   (2,3-Benzopyridine)   dicarboxylic acid

Pyridine can be reduced by reagents such as sodium-and-alcohol which have no effect on benzene. Alkyl groups attached to the pyridine nucleus can be oxidized to yield the corresponding pyridine-carboxylic acid. The generalization that pyridine undergoes addition more readily but is more resistant to oxidation than benzene is usually true; however like all generalizations there are exceptions and the reactions of pyridine depend a great deal on the conditions. Thus 2-phenylpyridine, on treatment with acidic permanganate, yields pyridine-2-carboxylic acid (picolinic acid) while treatment with alkaline permanganate yields benzoic acid.

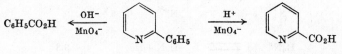

Benzoic acid   2-Phenylpyridine   Pyridine-2-carboxylic
                                                acid
                                                (Picolinic acid)

This follows a general trend; we shall see later that pyridine is much more susceptible to nucleophilic and basic reagents than benzene but much more resistant to electrophilic and acidic reagents.

The great distinguishing feature of the chemistry of benzene was its tendency to undergo addition-with-elimination reactions with electrophilic reagents. Weak electrophiles fail to attack the benzene nucleus while strong electrophiles add and a proton is then eliminated so that the overall reaction results in substitution (this includes the familiar reactions of nitration, sulphonation, etc.). However, we must remember that pyridine, besides being an aromatic compound, is also a tertiary base and a nucleophile. We have already seen that pyridine reacts with a proton to form the pyridinium ion ($C_5H_5NH^+$). Thus, whereas benzene reacts with nitronium fluoroborate to yield nitrobenzene, pyridine forms the salt nitropyridinium fluoroborate.

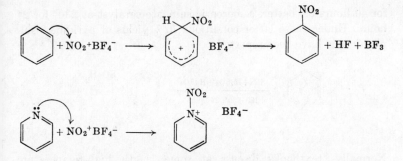

Similarly, benzene treated with sulphur trioxide undergoes sulphonation, whereas pyridine forms a stable complex. This complex is a sulphonating agent and has considerable application for the sulphonation of reactive aromatic compounds.

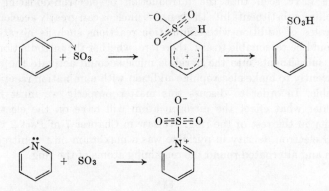

It is possible to nitrate trimethylanilinium salts, so we might reasonably ask whether it is possible to nitrate pyridinium salts. The answer to this question is 'No'. The positive charge in the trimethylanilinium salt is on an atom adjacent to the ring whereas the positive charge in the pyridinium salt is on an atom actually forming part of the ring. Some textbooks state that pyridine can be nitrated by reaction with sulphuric acid and sodium nitrate at 330°, 3-nitropyridine being said to be formed in 20% yield. In fact, subsequent workers were unable to repeat this and report instead that only 6% of a mixture of nitropyridines is formed. Sulphonation can be achieved under extreme forcing conditions, by using either oleum at 350°

for 40 hours or, better, a mercuric sulphate catalyst at 230° for 24 hours. Under these latter conditions 70% yields of pyridine-3-sulphonic acid have been reported.

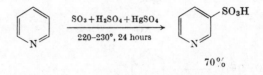

$$\text{SO}_3 + \text{H}_2\text{SO}_4 + \text{HgSO}_4$$
220–230°, 24 hours

70%

Normally, textbooks discuss at some length high-temperature bromination and chlorination of pyridine. In fact, none of these reactions proceeds very smoothly and for practical purposes it is better to say that pyridine does not undergo electrophilic addition-with-elimination reactions apart from sulphonation and mercuration.

We have seen that the introduction of electron-donating or -repelling substituents into the benzene nucleus can greatly accelerate the rates of addition-with-elimination reactions such as nitration. It would be reasonable to ask, therefore, whether the introduction of such substituents into the pyridine nucleus could activate the ring sufficiently to make electrophilic addition-with-elimination reactions possible. In order to discuss this matter properly we must first consider what effect the nitrogen atom will have on the electron density in the rest of the ring. We saw in Chapter 7 of Part 2 that the π electron density in pyridine was a maximum on the nitrogen atom and alternated round the remaining atoms of the ring.

π Electron density in pyridine
(see Part 2, page 61)

In the pyridinium ion this alternation of charge is even more marked. The presence of a positive charge will mean, however, that the whole ring is resistant to attack by an electrophile but that if such an attack occurs at all the positively charged nitrogen atom will tend to direct substitution to the 3-position. 2,4,6-Trimethyl-pyridine (symmetrical collidine), with three electron-repelling groups, can be nitrated to give the 3-nitro-compound in high yield.

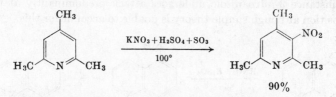

A single donor group activates the pyridine ring sufficiently to permit nitration and halogenation. Thus 4-aminopyridine can be halogenated, sulphonated, and nitrated to give the corresponding 3-substituted and then the 3,5-disubstituted 4-aminopyridines.

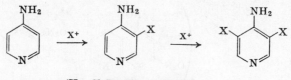

$$(X = Cl, Br, NO_2, \text{ or } SO_3H)$$

Similarly, 2-aminopyridine undergoes electrophilic addition-with-elimination reactions, the substituent appearing first in the 5-position and then on further reaction in the 3-position. So long as the donor group is in the 2- or 4-position there is no problem about directive effects since the positively charged nitrogen (in all these reactions the first step will be the formation of the quaternary pyridinium derivative) and the donor group both direct the attack of the electrophile into this 3- or 5-position. When the donor group is in the 3-position, however, there is a conflict of directive effects, but the positively charged nitrogen is only an electron attractor, even though it is a very powerful one (i.e. can only exert an inductive I_π effect). The donor group, therefore, determines the site of attack (see Chapter 8 of Part 2). Thus 3-hydroxypyridine is readily nitrated by a mixture of nitric acid and sulphuric acid to yield 2-nitro-3-hydroxypyridine.

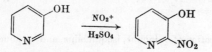

Attack at the 2-position, i.e. between the donor and the attractor groups, rather than the 4- or 6-position is a quite general phenomenon; for instance 3-nitroanisole undergoes attack predominantly at the 2-position although simple theory is unable to account for this.

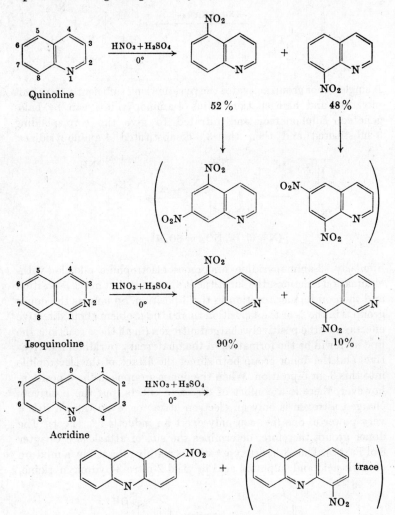

Nitration of quinoline, isoquinoline, and acridine

It is worth briefly considering the electrophilic addition-with-elimination reactions of the benzopyridines. These compounds are heterocyclic analogues of naphthalene and anthracene. From all that we have said about the chemistry of pyridine we should expect any attack by an electrophile to occur in the benzene ring of these compounds. By analogy with naphthalene we would expect the attack in quinoline and isoquinoline to occur predominantly at the 5- and the 8-position, which is what is observed (see p. 64).

Notice first that substitution occurs in the non-heterocyclic ring; secondly, that the attack occurs where we would have predicted for naphthalene in the cases of quinoline and isoquinoline; and thirdly, that further substitution occurs in the non-heterocyclic ring in spite of the fact that it is already substituted by a nitro group.

Neither pyridine nor its benzo-derivatives are attacked by weak electrophiles such as those involved in Friedel–Crafts acylation and alkylation or in nitrosation.

Nucleophiles do not add to benzene, but we saw in Chapter 14 of Part 2 that if there are sufficient electron-attracting groups, or better, electron-accepting groups attached to the benzene nucleus nucleophilic addition is possible. Thus 1,2-dichloro-3,5-dinitrobenzene reacts extremely rapidly with common nucleophiles such as the hydroxide anion or the methoxide anion or even such weak nucleophiles as primary and secondary amines.

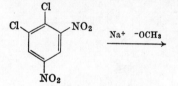

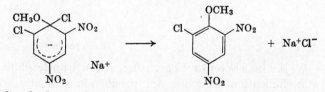

We have already drawn the conclusion that the heterocyclic nitrogen atom can only exert an inductive effect (as an electron-attractor). It is far more effective in this capacity than even a positively charged quaternary nitrogen atom attached to a benzene ring. We would conclude, therefore, that pyridine would be susceptible to nucleophilic attack at the 2- and the 4-position, and this is what is found.

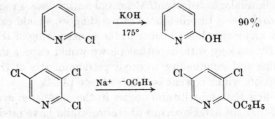

Besides the hydroxide and alkoxide ions, all the usual nucleophiles are effective. For example, 2-bromopyridine, treated with concentrated aqueous ammonia solution in a sealed tube, gives high yields of 2-aminopyridine. Other common nucleophiles which are effective include the hydrogen sulphide ion and primary and secondary amines. As with benzene the reactivity of the group being replaced is in the order $F > Br > Cl > NO_2$. Halogens in the 3-position are not replaced by ordinary nucleophiles but treatment with strong bases such as sodamide results in replacement, not simply to give the 3-amino-compound, but a mixture of the 3- and the 4-amino-compounds; it seems probable, therefore, that under these conditions dehydro-pyridine is formed (cf. Chapter 20 of Part 2).

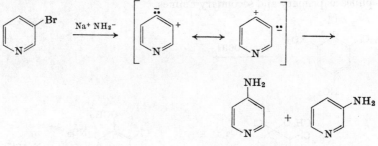

(main product)

We saw in Chapter 13 of Part 2 that, although nucleophilic addition-with-elimination reactions are possible in aromatic compounds containing sufficient acceptor or attractor groups, the elimination step does not occur if the nucleophile attacked a site carrying a hydrogen atom. Thus, although an alkoxide anion might add to 2,4-dinitrobenzene, no replacement occurred because this would involve the ejection of a hydride ion, a process energetically highly unfavourable. At first sight there is no reason why pyridine should behave differently. In practice, however, it is found that hydrogen atoms in the positions 2 and 4 can be replaced by very powerful nucleophiles such as the amide anion and to some extent even by weaker nucleophiles such as the hydroxide anion.

The amination of pyridine and its derivatives by alkali-metal amides is often known as Tchitchibabin reaction. When pyridine is heated with sodamide in dimethylaniline solution it is converted into 2-aminopyridine in 60—70% yield.

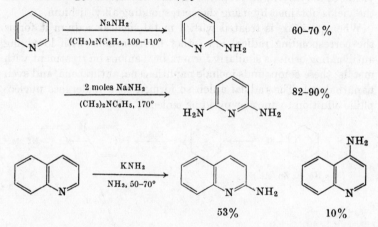

The amino group enters the 2- and then the 6-position. A small amount of 4-aminopyridine is formed together with traces of bi-pyridyls. 2-Methylpyridine (α-picoline), when treated with sodamide in liquid ammonia at low temperatures, yields the corresponding carbanion owing to abstraction of a proton from the methyl group; heating, however, leads to amination, yielding 2-amino-6-methyl-pyridine. Quinoline is aminated more readily than pyridine and a higher proportion of the product is the 4-amino-isomer; attack is restricted to the heterocyclic ring.

Very reactive carbanions will also add to the pyridine nucleus. If the initial adduct is treated with water the substituted 1,2-dihydro-pyridine is formed. Alternatively, if the intermediate is heated at 90—100° a hydride ion is eliminated and the alkylpyridine is formed directly.

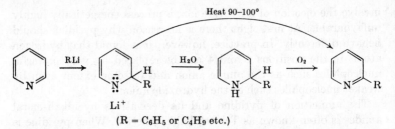

(R = C₆H₅ or C₄H₉ etc.)

Even Grignard reagents are sufficiently nucleophilic, although the
yields of alkylpyridine prepared in this way are much poorer than
the yields obtained by using the corresponding alkyl-lithium.

When pyridine is treated with a metal such as sodium it forms
the corresponding radical ion (other hydrocarbons with low-lying
antibonding orbitals similarly form radical anions on treatment with
metals; these compounds include naphthacene, anthracene, and even
naphthalene). The radical anion so formed then undergoes nucleophilic addition to another pyridine molecule.

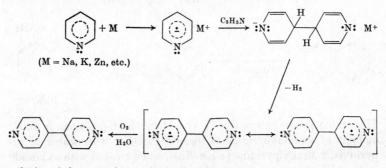

(broken circles are used to emphasize the delocalization of the odd electron in the
radical anions)

When pyridine is treated with sodium at moderately elevated temperatures a mixture of all possible bipyridyls is formed, although
4,4'-bipyridyl remains a predominant product. Similar reactions
occur with zinc and magnesium; thus when pyridine is treated with
zinc and acetic anhydride the intermediate radical anions react
with the acetic anhydride to form 1,1'-diacetyl-4,4'-dihydro-4,4'-
bipyridyl. This compound disproportionates on being heated to give
pyridine and 1,4-diacetyl-1,4-dihydropyridine which can in turn, be
converted into 4-ethylpyridine by treatment with zinc and acetic acid.

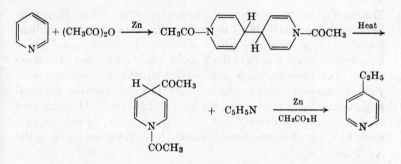

Derivatives of pyridine can be divided into two classes according to whether the substituent is attached to a carbon atom or to the nitrogen atom. We shall consider the former first. In general, these compounds undergo the same reactions as other aromatic compounds having such substituents. Halogenopyridines, besides undergoing nucleophilic displacement reactions discussed above, form normal Grignard reagents although rather special conditions are necessary. Similarly, the lithium derivatives can be formed from the halogen derivatives by treating the 2- or 3-bromopyridine with butyllithium at low temperatures. Just as toluene can be oxidized to benzoic acid, so the methylpyridines can be oxidized to the corresponding pyridinecarboxylic acids. A 2- or 4-methyl group is in a site analogous to the methyl group in *p*-nitrotoluene and, as we have mentioned briefly above, a carbanion can be formed by removal of a proton when the alkylpyridine is treated with strong base. The resulting carbanion undergoes all the reactions we should expect.

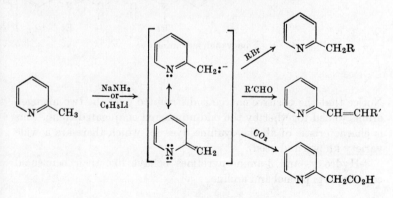

The chart shows the reactions of 2-methylpyridine; 4-methylpyridine behaves very similarly but 3-methylpyridine does not form a carbanion of this type anywhere nearly as readily. If the nitrogen atom is quaternized, alkyl groups in the 2- and 4-positions become even more acidic. The reaction of 2- and 4-alkylquinolinium derivatives to yield compounds called cyanine dyes has considerable practical importance. An example of such a dye is Pinacyanol. This compound is of no value as a dyestuff but it has the property of sensitizing a silver bromide photographic emulsion to light in the red region of the spectrum.

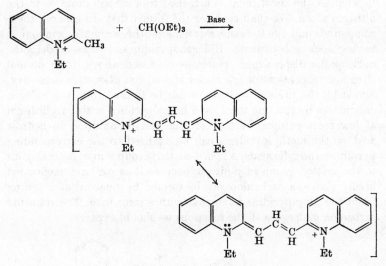

Pinacyanol, **a** cyanine dye

Notice that the positive charge is distributed over the two nitrogen atoms joined together by the odd-numbered conjugated chain. This is characteristic of all the cyanine dyes (of which there are a wide variety known and used).

3-Hydroxy- and 3-amino-pyridines behave like their benzenoid counterparts phenol and aniline.

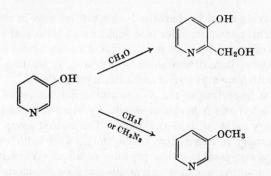

When we come to consider 2- and 4-hydroxypyridines and amino-pyridines a complicating factor arises. 4-Hydroxypyridines can tautomerize to yield 4-pyridone.

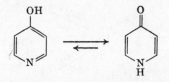

There is a great deal of evidence (from IR, UV, and NMR spectroscopy) to show that both 2- and 4-hydroxypyridine exists almost entirely in the pyridone form both in solution and in the solid state. Methylation with methyl iodide or dimethyl sulphate yields predominantly the *N*-substituted product whereas methylation with diazomethane in the case of 2-pyridone yields the *O*-methyl derivative.

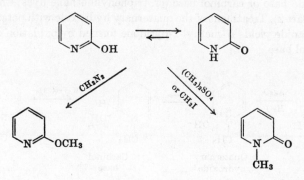

3-Aminopyridine can be diazotized with nitrous acid in the normal way to yield diazonium salts that undergo all the usual reactions. 2- and 4-Aminopyridines are not diazotized under normal conditions although the sodium diazoate can be prepared by treating 2-amino-pyridine with isopentyl nitrite and sodium ethoxide.

The most important of the *N*-substituted derivatives of pyridine are the *N*-alkyl and *N*-aryl compounds. Pyridine very readily forms a quaternary iodide with methyl iodide. The methyl group attached to a quaternary nitrogen atom readily forms a carbanion and adds to carbonyl compounds in the presence of alkali. When heated at 300° the quaternary iodide rearranges to give moderate to poor yields of the 2- and 4-methylpyridinium iodide.

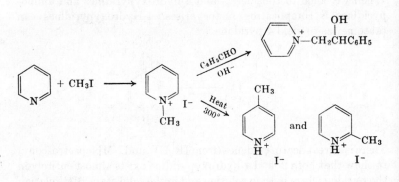

The quaternary iodide can be converted into quaternary hydroxide in the usual way. This compound appears to be in equilibrium with a pseudo-base or carbinol base (cf. triphenylmethane dyes; Chapter 19 of Part 2). Treatment of the quaternary hydroxide with potassium ferricyanide yields *N*-methyl-2-pyridone formed by oxidation of the carbinol base.

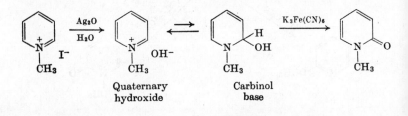

Quaternary hydroxide   Carbinol base

With pyridine derivatives the equilibrium lies on the side of the quaternary hydroxide but *N*-methylquinolinium hydroxide exists almost entirely as the carbinol base which can tautomerize further to the open-chain form.

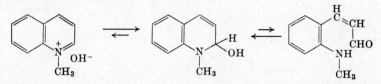

This kind of ring cleavage can be facilitated in pyridine itself when an electron-attracting group is attached to the nitrogen. Thus 1-chloro-2,4-dinitrobenzene quaternizes readily with pyridine and the heterocyclic ring opens in the presence of base. We already referred to this reaction, which yields derivatives of glutaconic aldehyde, in our discussion of the synthesis of azulene in the preceding chapter (p. 42).

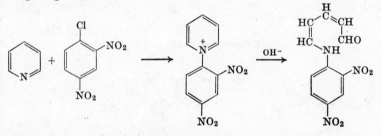

Acid chlorides form quaternary acyl compounds with pyridine and its derivatives. A solution of benzoyl chloride in pyridine is a very effective benzoylating agent. Characteristically quinoline compounds undergo addition to the ring more readily than the pyridinium compound. Thus the salt from benzoyl chloride and quinoline adds nucleophiles very readily. The compound formed by the addition of the cyanide ion is known as a Reissert compound, and on treatment with acid it hydrolyses to yield benzaldehyde and quinoline-2-carboxylic acid. This reaction is sometimes used for preparing aldehydes from acid chlorides when other methods are unavailable.

The exact mechanism of the formation of the aldehyde is unknown, although there is evidence that a cyclic intermediate is involved, and a possible sequence is depicted at the bottom of p. 74

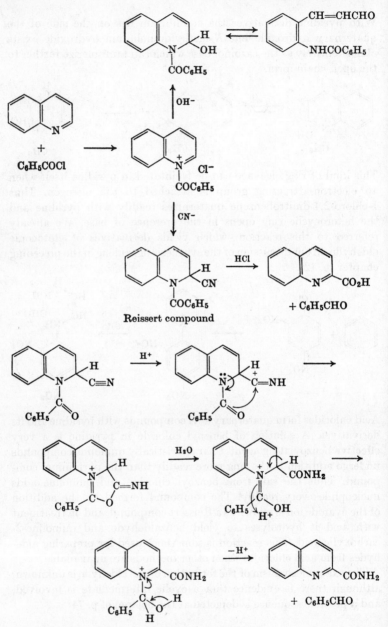

Reissert compound

The remaining important *N*-substituted derivative of pyridine we have to consider is the oxide. Tertiary amines treated with a peracid (e.g. hydrogen peroxide in acetic acid) yield *N*-oxides and pyridine is no exception.

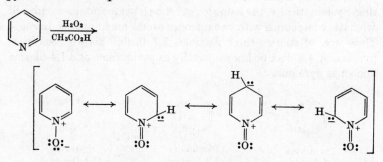

The important feature of this molecule is that the oxygen atom has filled *p* orbitals, one of which will have the same symmetry as the π orbitals of the pyridine ring, so that there is electron delocalization over the whole π electron system. Effectively, the oxygen atom will behave as a donor. Pyridine *N*-oxide is thus a weaker base than pyridine itself, and provided that the solution is not too acidic it will undergo electrophilic addition-with-elimination reactions, predominantly in the 4-position. For example, nitration of pyridine *N*-oxide gives a high yield of the 4-nitro-compound.

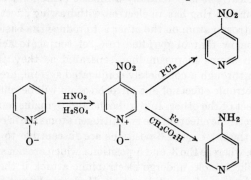

In very strong acid the oxide becomes protonated and under these circumstances substitution will occur, if at all, in the 3-position (this happens when sulphonation is attempted using oleum).

Diazines, Triazines, and Tetrazines

We have regarded pyridine as a benzene molecule in which a nitrogen atom has replaced a carbon and a hydrogen atom. No new principle is involved in introducing further nitrogen atoms into the ring. Systematically, the name for pyridine is azine and six-membered aromatic compounds with two nitrogen atoms are known as diazines. There are, of course, three diazines, 1,2-diazine known usually as pyridazine, 1,3-diazine known usually as pyrimidine, and 1,4-diazine known as pyrazine.

Pyridazine
(1,2-Diazine)

Pyrimidine
(1,3-Diazine)

Pyrazine
(1,4-Diazine)

The chemistry of these compounds can almost entirely be predicted from what we already know about the chemistry of benzene and pyridine. Pyridazine and pyrazine are not of great general importance. On the other hand, the pyrimidine nucleus occurs very widely in important natural products and its chemistry is therefore of considerable interest.

Pyrimidine is a low-melting colourless compound (m.p. 22.5°, b.p. 124°). Clearly it is potentially a dibasic compound but since a nitrogen atom in a ring has an electron-withdrawing effect the effect of the one nitrogen atom on the other is to reduce the basicity of the compound below that of pyridine (pK_a of conjugate acid = 1.30). The nitrogen atoms in pyrimidine, separated as they are by one carbon atom through a completely conjugated system, are placed so that any electronic effects of one nitrogen atom will be reinforced by the same effect of the other. In pyridazine and pyrazine, on the other hand, the electronic effects of the two nitrogen atoms are in opposition. In pyridine the 2-, 4-, and 6-positions are susceptible to attack by nucleophiles, whereas the 3- and 5-positions, which are not susceptible to nucleophilic attack, undergo electrophilic attack if the molecule can be attacked by electrophiles at all. Hydroxy groups and amino groups in the 2-, 4-, or 6-position behave abnormally because of their ability to tautomerize whereas the same substituents in the 3- and 5-positions behave as they do when attached to a benzenoid nucleus.

Examination of the pyrimidine nucleus shows that the 2-, 4-, and 6-positions should be similar to the corresponding positions in pyridine and position 5 should be analogous to the 3- and 5-positions in pyridine.

As we should expect, electrophiles do not add to the pyrimidine nucleus unless activating groups are present. Electrophilic addition-with-elimination reactions occur in pyrimidine nuclei activated by donor groups. Substitution occurs in the 5-position if free, again as expected.

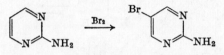

Pyrimidine with an amino group in the 5-position reacts with nitrous acid to give the expected diazonium salt, but interestingly 4- and 6-aminopyrimidines, treated with nitrous acid, undergo electrophilic addition-with-elimination at the 5-position if this position is vacant, rather than diazotization of the amino groups.

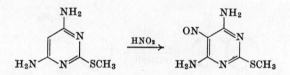

We should expect nucleophilic addition-with-elimination reactions to occur readily, and although unsubstituted pyrimidine has received little study, 6-methylpyrimidine undergoes amination, yielding a mixture of the 2- and 4-amino-6-methylpyrimidines.

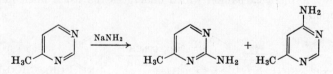

Naturally, nucleophilic addition-with-elimination of a halogen occurs even more readily provided that the halogen atom is in the 2-, 4-, or 6-position.

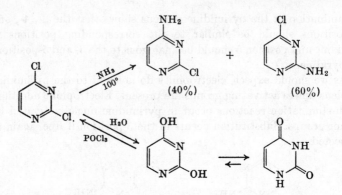

In pyrazine and pyridazine all the carbon atoms are activated to nucleophilic attack.

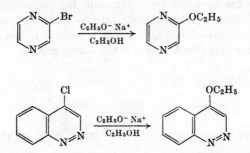

We have already commented on the fact that substituents in the 5-position in pyrimidine behave like those in benzene whereas those that are in the 2-, 4-, and 6-positions behave like substituents in the corresponding positions in pyridine. Thus, 2-methylpyrimidine readily forms a carbanion which undergoes the usual nucleophilic reactions.

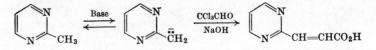

5-Hydroxypyrimidines behave like phenol whereas a pyrimidine containing a hydroxy group in the 2-, 4-, or 6-position exists entirely as the pyrimidone tautomer.

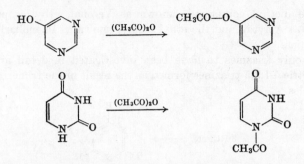

2,4-Dihydroxypyrimidine (common name uracil) is an important constituent of the nucleic acids (See Part 4). 2,4,6-Trihydroxypyrimidine is known as barbituric acid and is a stronger acid than acetic acid. This high acidity can be attributed in large measure to the symmetrical anion where the negative charge is evenly distributed between two of the oxygen atoms. Anything which inhibits this conjugation greatly reduces the acidity of the compound, so 5,5-diethylbarbituric acid and alloxan are both very much weaker acids.

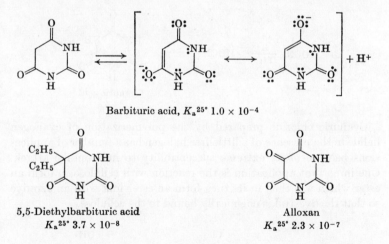

Barbituric acid, $K_a{}^{25°}$ 1.0 × 10⁻⁴

5,5-Diethylbarbituric acid
$K_a{}^{25°}$ 3.7 × 10⁻⁸

Alloxan
$K_a{}^{25°}$ 2.3 × 10⁻⁷

Derivatives of barbituric acid with two alkyl groups attached to carbon atom 5 are important as hypnotic and sedative drugs. The barbiturates are probably the most widely used and safest of the sedative drugs now used in medicine. Particularly important ones are the diethyl compound, known as Veronal, ethyl isopentyl, known as Amytal, and the ethyl phenyl known as Phenobarbitone or Luminal.

The only triazines to have been investigated in detail are the symmetrical 1,3,5-triazines formed as the result of the trimerization of nitriles.

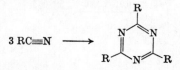

The parent compound prepared by the trimerization of hydrogen cyanide in the presence of an acid catalyst is extremely susceptible to attack by nucleophiles which results in complete decomposition of the ring. Treatment with water yields ammonium formate probably via formamidine, and treatment with amines leads to N-substituted formamidines. Cyanuric acid formed by the polymerization of cyanic acid appears to exist predominantly in the trione form.

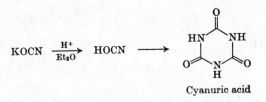

Cyanuric acid

Cyanuric chloride prepared by the polymerization of cyanogen halide in the presence of a little free halogen has a number of applications because of its extreme susceptibility to nucleophilic attack. One important application is the reaction with cellulose to form an ester which will then in its turn form an ester link with an azo dye so that the dyestuff is chemically bound to the cellulose.

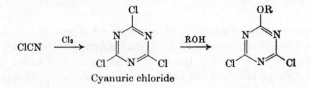

Cyanuric chloride

The only tetrazines known are the 1,2,4,5-derivatives. Symmetrical tetrazines are red to blue-red crystalline compounds almost non-basic and readily reduced to dihydro-compounds. The lack of basicity at one nitrogen can be attributed to the electron-withdrawing character of the others. The parent compound is unstable and is decomposed by the air although it can be kept in a sealed tube at low temperatures in the dark. 3,6-Diaryltetrazines are stable substances though they are decomposed by hot alkali and hot mineral acid. 3,6-Diphenyl-1,2,4,5-tetrazine undergoes a cycloaddition reaction with diphenylacetylene (tolan), the adduct losing nitrogen spontaneously to give 3,6-tetraphenylpyridazine. Heated by itself the diphenyltetrazine loses a molecule of nitrogen and yields two moles of benzonitrile.

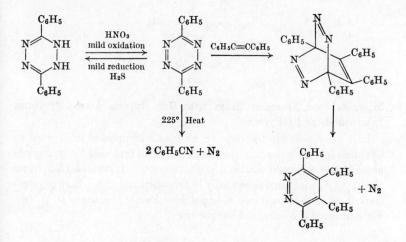

A biologically important bicyclic system is the pteridine nucleus. The parent compound, a yellow crystalline substance, is not very stable.

Pteridine
(yellow, m.p. 139°)

It adds water and when heated with aqueous sulphuric acid is degraded to 2-aminopyrazine-3-aldehyde. As we would expect from a compound formed by the fusion of a pyrimidine and a pyrazine ring, electrophilic substitution is unknown. Halogen atoms in any of the four available positions undergo nucleophilic substitution very readily.

Pteridines were first isolated as the yellow and white pigments of butterfly wings. However, their most important biological roles are as constituent parts of riboflavin (Vitamin B_2) and folic acid (also part of the Vitamin B complex).

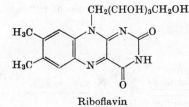

Riboflavin

Six-membered Aromatic Rings with One Oxygen Atom: Pyrylium Salts and 2- and 4-Pyrones

It is only possible to write an aromatic compound analogous to pyridine in which the nitrogen atom has been replaced by an oxygen atom if the molecule carries a positive charge. In such a compound the oxygen atom must necessarily be trivalent and this in turn requires that it must be positively charged (see Part 2, footnote to p. 66). We can draw the usual orbital picture:

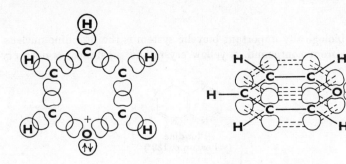

but in many respects the canonical forms of resonance theory give us a more useful picture.

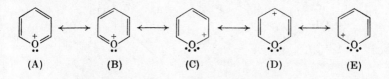

(A)   (B)   (C)   (D)   (E)

(C), (D), and (E) make much less contribution to the overall electronic distribution of the molecule than (A) and (B).

Pyrylium salts are stable crystalline substances. The parent compound is formed when the sodium salt of glutaconic aldehyde is treated with perchloric acid; the resultant pyrylium perchlorate is a colourless salt decomposing at 275°. In acidic media the pyrylium salts are extremely stable and 2,4,6-triphenylpyrylium perchlorate can be nitrated with fuming nitric acid to yield 2,6-di-*m*-nitrophenyl-4-*p*-nitrophenylpyrylium perchlorate.

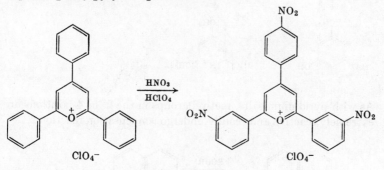

We should expect the main reactions of a pyrylium salt to be those typical of an oxonium ion. The addition of a nucleophile invariably results in ring opening but often the ring-opened compound cyclizes to yield a new aromatic compound (cf. the chart, in which the reactions of 2,4,6-trimethylpyrylium salts with nucleophiles are illustrated). Reaction with ammonia to yield a pyridine derivative is completely general. The reactions with hydroxide and with secondary amines depend on the fact that an activated methyl group is available in the ring-opened intermediate.

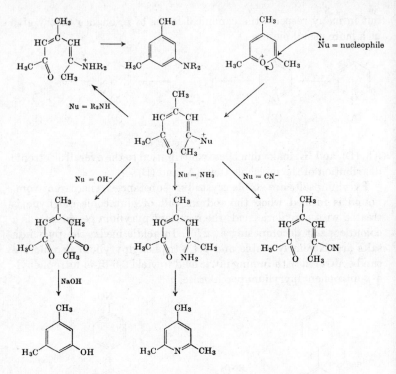

As with pyridinium salts, methyl groups in the 2- or 4-positions are extremely reactive and readily undergo condensation reactions.

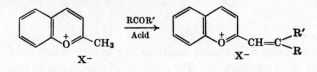

Polyhydroxy-derivatives of benzopyrylium, or more particularly 2-phenylbenzopyrylium, form the principal red or blue colouring materials of plants, providing the reds and blues of petals and the red and orange colours of autumn foliage. The pyrylium salts called anthocyanidins occur combined with sugar molecules, the complete natural products being called anthocyanins.

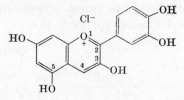

Cyanidin

(red colour of rose and poppy due to 3,5-diglucoside)

The types of sugar and their positions and number, vary from plant to plant, and in some plants the hydroxy groups attached to the 2-phenyl group are methylated or replaced by methylenedioxy groups. The biosynthesis of this group of compounds is discussed in Part 4.

In the previous chapter we saw that cycloheptatrienone (tropone) showed few of the reactions expected of a ketone, and we attributed this to a dipolar structure in which the oxygen atom accepted a pair of electrons from the ring and became negatively charged, leaving the seven-membered ring with six π electrons and a positive charge. 4-Pyrone (γ-pyrone) is an analogous compound.

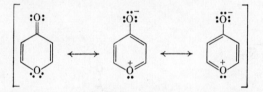

The contribution of the dipolar structures to the ground state of the molecule is fairly small. Thus, for example 4-pyrone has a dipole moment of approximately 4 D instead of nearly 22 D which would be expected for the dipolar structure. On the other hand the dipole moment of the ketonic structure would be less than 2 D. Similarly, the infrared spectrum of 4-pyrone shows a carbonyl band but its frequency is exceptionally low ($\nu = 1630$ cm^{-1}). Like tropone, 4-pyrone forms stable salts with acids.

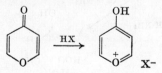

Similarly, 2,6-dimethyl-4-pyrone reacts with dimethyl sulphate to yield the 2,6-dimethyl-4-methoxypyrylium salt.

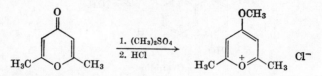

Under special conditions 4-pyrone will undergo electrophilic addition-with-elimination reactions, probably via the pyrylium salt.

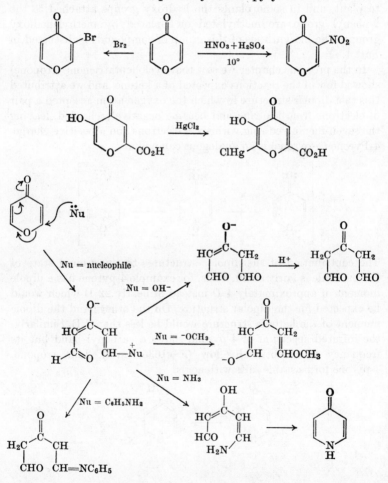

Like tropone, 4-pyrone shows few of the reactions of a ketone. Instead, nucleophiles add to the 2-position with concurrent ring fission, as in reactions of pyrylium salts (see p. 86).

Although normally written as an unsaturated highly conjugated ketone 4-pyrone does not react with dienes.

Just as 2-phenylbenzopyrylium salts occur widely in plants, forming important colouring matters, so 2-phenylbenzopyrones (called flavones) occur widely in the plant kingdom. They are often yellow compounds and account for the yellow colours of many

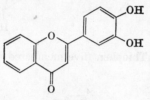

Luteolin
(orange-yellow; used as a natural dye)

petals. 3-Phenylbenzopyrones (called isoflavones) also occur widely in the plant kingdom and some of these have physiological action.

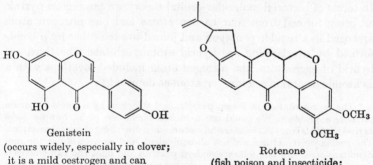

Genistein
(occurs widely, especially in clover;
it is a mild oestrogen and can
cause miscarriages in ewes)

Rotenone
(fish poison and insecticide;
occurs in *Derris* root)

The very interesting biosynthetic relations of these compounds are discussed in Part 4.

2-Pyrones (α-pyrones) can be written with canonical forms analogous to those of 4-pyrone. In practice, however, 2-pyrone and its derivatives behave as unsaturated aliphatic compounds, undergoing Diels–Alder reactions, polymerizing on standing, and in general behaving like the lactones of unsaturated carboxylic acids.

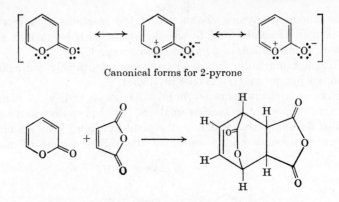

Canonical forms for 2-pyrone

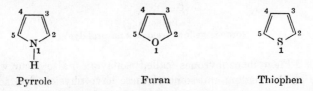

Pyrrole, Furan, and Thiophen (five-membered rings containing one heteroatom)*

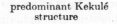

Pyrrole   Furan   Thiophen

In terms of pictorial molecular-orbital theory we can regard pyrrole as being formed from four carbon atoms and one nitrogen atom arranged in a regular pentagon and joined to each other by σ bonds formed by overlapping sp^2 hybrid atomic orbitals. The third sp^2 hybrid of each atom, the nitrogen atom included, overlaps with a $1s$ atomic orbital of a hydrogen atom as depicted.

* The representation of furan, pyrrole, and thiophen by a single structure presents a problem. We could use a broken circle (see p. 8), but we have agreed that this is not the most useful picture when discussing chemical reactions. The resonance structures are not all equivalent, and so we will use the predominant structure but follow standard practice and omit the two electrons donated to the π electron system by the heteroatom, unless they are involved in the reaction we are discussing

'broken circle'   predominant Kekulé   abbreviated structure
representation   structure

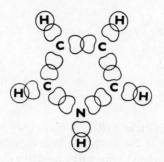

This leaves a p_z atomic orbital associated with each carbon atom and with the nitrogen atom which can overlap to form five π molecular orbitals. Each carbon atom will contribute one electron to this π orbital system while the nitrogen atom will contribute two. The molecule we so obtain is isoelectronic with the cyclopentadienyl anion. Remember that in that anion the lowest-energy bonding orbital is non-degenerate and the next two orbitals occur in pairs, ψ_2 and ψ_3 bonding and degenerate, ψ_4 and ψ_5 antibonding and degenerate. The presence of the heteroatom disturbs the relative energies of the orbitals, so that they no longer occur in degenerate pairs; effectively the electronegative nature of the nitrogen atom tends to lower the energies of ψ_2 and ψ_4. The really important point, however, is that the first three π orbitals of pyrrole are all bonding so that all six electrons go into bonding orbitals.

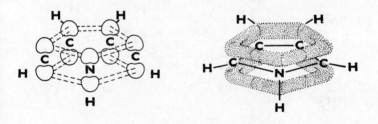

In terms of resonance theory there is only one non-ionic Kekulé structure although there are four dipolar structures with a positive charge on the nitrogen and negative charges on the carbon atoms.

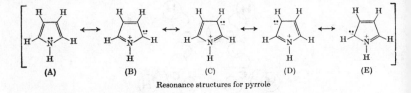

Resonance structures for pyrrole

Structures (B), (C), (D), and (E) will make only a small contribution to the ground state of the molecule. It is significant, however, that in all of them the carbon atoms carry negative charges.

At the beginning of the chapter we regarded pyridine as benzene in which one carbon atom had been replaced by a nitrogen atom. Nitrogen contains one more electron than carbon and instead of forming an exocyclic bond with a hydrogen atom the nitrogen atom in pyridine has a lone pair of electrons in an sp^2 hybrid orbital projecting out of the ring. We can regard furan as being related to pyrrole in the same way as pyridine is related to benzene. The four carbon atoms and the oxygen atom are arranged at the corners of a pentagon and form σ bonds by overlapping sp^2 hybrid orbitals. The remaining hybrid orbital of each carbon atom overlaps with a $1s$ atomic orbital of the hydrogen atom while the third sp^2 hybrid orbital of the oxygen atom contains two electrons forming a non-bonding lone pair. The $2p_z$ atomic orbitals on the carbon atom and the oxygen atom overlap to form five π orbitals exactly analogous to those in pyrrole.

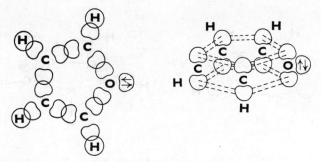

As with pyrrole we can only draw one Kekulé structure which is uncharged, the remaining structures being dipolar.

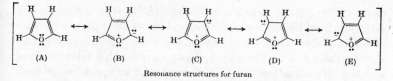

Resonance structures for furan

Thiophen could be represented in exactly the same way as furan, but there are complications associated with the fact that the valency electrons of sulphur have the principal quantum number 3 and hence in addition to *s* and *p* orbitals, *d* orbitals are also available for bonding. The discussion of this must wait until Chapter 5 and we will simply content ourselves here with resonance pictures. We are familar with the idea that sulphur can be bivalent like oxygen, but it also can be tetravalent and hexavalent. This allows us to draw additional canonical forms.

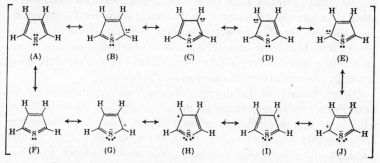

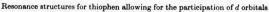

Resonance structures for thiophen allowing for the participation of *d* orbitals

Structures (A) to (E) are analogous to those we had for pyrrole and furan, but in structures (F) to (J) the sulphur atom has 10 electrons in its outer shell, not 8 as in the other structures.

As often happens the resonance pictures tell us more than pictorial molecular-orbital theory. If our picture is correct, then in each of these heterocyclic molecules the heteroatom is to some extent donating electrons to the delocalized system of electrons of the ring. In pyrrole, for example, the nitrogen atom has 8 electrons in its outer shell, 6 of which are shared with two carbon atoms and a hydrogen atom to form the single bonds, the remaining 2 being shared with 4 carbon atoms. If the delocalization of the electrons in the π orbitals were complete the nitrogen atom would carry 4/5 of a positive charge and each carbon atom would carry 1/5 of a negative charge. This charge distribution is depicted in structures (B), (C), (D), and (E) of the resonance forms; and the same applies in the case of furan. Since nitrogen and oxygen are more electronegative than carbon, there will be a strong tendency to resist this delocalization, and the final situation will be the result of a compromise between the energy gained by delocalizing the electrons and the energy lost by creating a dipole. Oxygen is more electronegative than nitrogen (see Chapter 5 of Part 2), and we are familiar with the fact that an oxygen atom in a compound such as water is far less ready to donate its pair of non-bonding electrons than nitrogen is in a trivalent compound such as ammonia (ammonia is a strong base, water a weak one). Thus we should expect electron delocalization to be less import-ant in furan than in pyrrole, or, putting it another way, we should expect furan to be less 'aromatic' than pyrrole. Conversely, we should expect, furan to show more of the properties of an unsaturated ether than pyrrole will show of an unsaturated amine. In our reson-ance pictures for thiophene the sulphur both donates and accepts electrons, so that there should be no net charge on the sulphur atom. In practice, there is very good evidence that electron delocalization is almost as complete in thiophen as it is in benzene.

The reactions that distinguish benzene from hypothetical cyclo-hexatriene are its reluctance to undergo addition, its stability to oxidation, and the fact that when addition does occur the initial adduct ejects a proton regenerating the aromatic nucleus in the addition-with-elimination reactions. We shall consider to what extent furan, pyrrole, and thiophene exhibit 'aromatic' properties. We shall always deal with these compounds in this order, namely, furan, pyrrole, and thiophen, because we have seen from the arguments above that we should expect furan to be the least and thiophen to be the most aromatic. All three compounds are more readily reduced than benzene, although direct comparison is rather difficult. Furan and pyrrole can be reduced to tetrahydrofuran and pyrrolidine (tetrahydropyrrole) respectively, by treatment with hydrogen and Raney nickel under pressure.

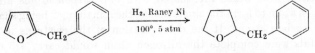

2-Benzylfuran   2-Benzyltetrahydrofuran

Notice that under the conditions specified the benzene ring is un-affected. Similarly, pyrrole can be reduced to its dihydro-derivative (pyrroline) by zinc dust and acetic acid, a reagent that has no effect on benzene.

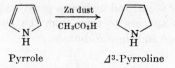

Pyrrole Δ^3-Pyrroline

Thiophen, when treated with Raney nickel, invariably undergoes desulphuration until all the nickel catalyst is poisoned, a reaction of diagnostic and synthetic importance.

$$\text{(thiophen)} \xrightarrow[\text{[H}_2\text{]}]{\text{Raney Ni}} H_3C \begin{array}{c} CH_2-CH_2 \\ \diagdown CH_3 \end{array} + \text{NiS}$$

Naphthalene is more readily reduced than benzene and likewise the benzo-derivatives of the five-membered heterocyclic compounds, benzofuran, indole, and thionaphthen are all reduced more easily than the parent compounds by, for instance, sodium-and-alcohol. It is significant, however, that under these conditions the reduction takes place exclusively in the heterocyclic ring.

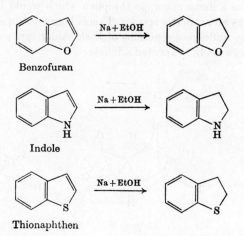

Benzofuran   Na + EtOH

Indole   Na + EtOH

Thionaphthen   Na + EtOH

All three heterocyclic nuclei are more readily oxidized than benzene
though a lot depends on the reagents and the conditions. In general,
the products are complex and the result of extensive degradation.
With thiophen, however, there is the additional possibility of oxidation of the sulphur atom, and hydrogen peroxide in acetic acid gives
a compound that appears to arise from the Diels–Alder addition
between the S-oxide and the S,S-dioxide.

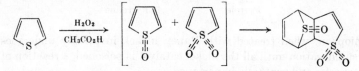

The thiophen nucleus is, however, substantially the most stable and
methylthiophen can be oxidized with chromic acid to the corresponding 2-carboxylic acid, a reaction that is not possible with 2-methyl-
furan or 2-methylpyrrole.

On the basis of our initial discussion we should expect these
five-membered heterocyclic compounds to show both the reactions
of conjugated dienes and those of aromatic nuclei. We should expect
a transition of properties from furan, which would behave predominantly as a diene ether, to thiophen which would behave predominantly as an aromatic species. This is borne out in experiment.
Furan readily undergoes reactions of Diels–Alder type; with maleic
anhydride it yields the expected adducts.

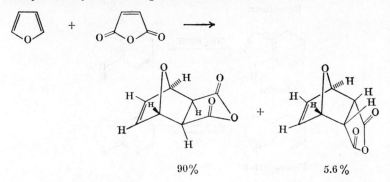

90%   5.6%

In Chapter 21 of Part 2 the reaction of furan with dehydrobenzene
was described.

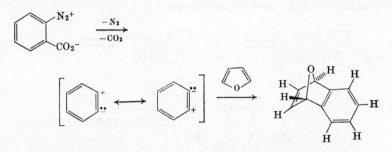

1-Methylpyrrole gives a very poor yield of adduct with dehydrobenzene, and although pyrrole it reacts with maleic anhydride to give
2-pyrrolylsuccinic acid this is quite different from a cycloaddition
reaction. Thiophen undergoes none of these addition reactions.

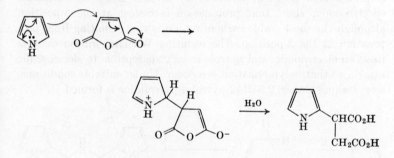

When an electrophile adds to an olefin the resultant carbonium
ion combines with a nucleophile so that the overall reaction is one
of addition to the double bond. When benzene reacts with an electrophile, however, the resultant positively charged Wheland intermediate
normally loses a proton to regenerate the resonance-stabilized aromatic system. Furan is on the border line. It behaves both as a
conjugated diene undergoing addition reactions and as an aromatic
compound undergoing addition-with-elimination reactions. The first
electrophile we must consider is a proton; furan has a lone pair of
electrons available on the oxygen atom and we should expect protonation at this site. The resulting oxonium ion is very unstable, and

treatment of furan with aqueous hydrochloric acid yields a mixture of succindialdehyde and polymer. The ring opening reaction can be quantitative in 2,5-disubstituted furans.

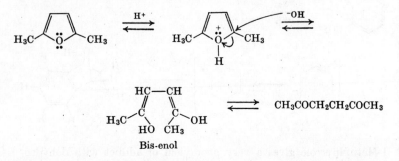

Bis-enol

In pyrrole no lone pair of electrons is available and protonation occurs at one of the carbon atoms. Exchange reactions with deutero-sulphuric acid (D_2SO_4), observed by nuclear magnetic resonance spectroscopy, show that protonation is fastest at the 3-position although the most stable carbonium ion is that resulting from pro-tonation at the 2-position. The resulting Wheland intermediate is itself an electrophile, and pyrrole is very susceptible to electrophilic attack, so that polymerization can occur. Under suitable conditions, a crystalline trimer, 2,5-di-(2-pyrrolyl)pyrrolidine is formed.

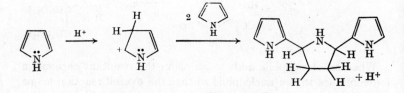

Thiophen is reasonably stable in acid solution although it is decomposed in concentrated sulphuric acid solution, polymeric substances being formed.

After considering the addition of a proton the next electrophiles we normally consider are bromine cations. Ionic bromination vividly distinguishes the reactivities of the three molecules. Furan normally gives a complex mixture of products but under special conditions products of 2,5-addition can be isolated. 2-Furoic acid (furan-2-

carboxylic acid) treated with bromine at −15° yields the saturated tetrabromo-compound as a result of addition. Pyrrole undergoes electrophilic addition-with-elimination reactions extremely readily and in the presence of an excess of bromine in alcohol the tetrabromo-pyrrole is formed. Thiophen is much less reactive and bromination in acetic acid at 0° yields a mixture of 2-bromothiophen and 2,5-dibromothiophen, the former predominating.

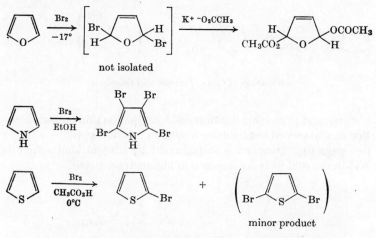

Bromination of furan, pyrrole, and thiophen

It would be incorrect to conclude that furan derivatives never undergo electrophilic addition-with-elimination reactions with the halogens. In fact, at higher temperatures 2-furoic acid yields a mixture of 4- and 5-bromo-compounds. The situation on nitration is very similar. Treated with nitric acid and acetic anhydride at low temperatures furan undergoes 2,5-addition, characteristic of a diene, but in the presence of a base this adduct is readily converted into 2-nitrofuran. Pyrrole is decomposed by ordinary acidic nitration media, but 2-nitropyrrole can be formed by using non-acidic reagents. Thiophen is reasonably stable to acidic reagents and can be nitrated by a solution of nitric acid and acetic acid to give predominantly the 2-nitro-compound although 5% of the 3-nitro-compound is formed in the same reaction.

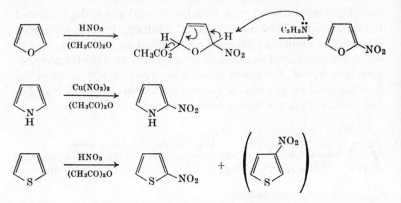

Nitration of furan, pyrrole, and thiophen

Furan and pyrrole are decomposed by sulphuric acid but sulphonation can be carried out by using a sulphur trioxide–pyridine complex (see page 61). Thiophen is sulphonated by concentrated sulphuric acid in the cold (it is decomposed in hot acid; see p. 96).

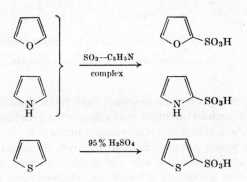

Sulphonation of furan, pyrrole, and thiophen

Furan and pyrrole are so reactive that acylation can be achieved by acetic anhydride without the presence of catalyst, although better yields are obtained in the presence of Lewis acids. Thiophen, being slightly less reactive, can be acylated with acetyl chloride using stannic chloride as a catalyst. Friedel–Crafts alkylation is not a practicable reaction with unsubstituted furan or pyrrole, but thiophen can be alkylated by, for example, isobutene and sulphuric acid.

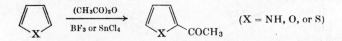

Friedel–Crafts acylation of furan, pyrrole, and thiophen

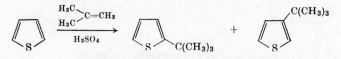

Friedel–Crafts alkylation of thiophen (only)

Before considering other electrophilic addition-with-elimination reactions it would be appropriate to summarize the results we have discussed so far. Furan shows only a limited aromatic character, and treated with electrophilic reagents often undergoes 2,5-addition characteristic of a conjugated diene rather than addition-with-elimination reactions characteristic of an aromatic nucleus. Pyrrole is by far the most reactive of the three compounds; we have drawn an analogy between furan and pyrrole on the one hand with pyridine and benzene on the other, so that we should expect furan to be far less reactive than pyrrole when it does undergo a true addition-with-elimination reaction. Very roughly we can compare the reactivity of pyrrole, furan, and thiophen with the activated benzenoid compounds resorcinol, phenol, and anisole respectively. Keeping this analogy in mind we should expect pyrrole to undergo all those electrophilic addition-with-elimination reactions characteristic of the poly-hydroxybenzenes such as the Gattermann–Koch reaction (see Part 2, p. 214), the Hoesch reaction (see Part 2, p. 215), coupling with diazonium salts, nitrosation, and iodination. The reactivity of the pyrrole nucleus is such that polysubstitution tends to occur particularly with the last three reactions although perfectly controlled monosubstituted products can be obtained if the pyrrole nucleus already contains electron-withdrawing substituents.

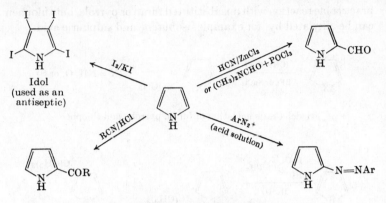

Reactions of pyrrole analogous to resorcinol

With furan the yields of aldehyde from the Gattermann reaction are appreciably lower than with pyrrole and neither iodination nor diazo coupling occurs. Thiophen is even less reactive and is unaffected by hydrogen cyanide and hydrogen chloride in the conditions of the Gattermann reaction although it undergoes chloromethylation (see Part 2, p. 213).

In all the reactions depicted we have shown the substituent as entering the 2-position in the heterocyclic nucleus. There are three possible resonance structures for electrophilic attack on pyrrole at the 2-position but only two for attack at the 3-position.

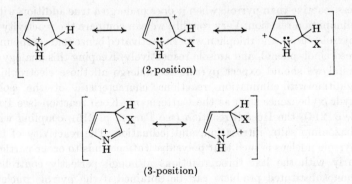

'Wheland intermediates' for electrophilic addition-with-elimination reactions

We have warned readers against deducing too much from the number of Kekulé structures, but what is significant in this case, as with ordinary aromatic substitution (see Chapter 8 of Part 2), is that the charge is more spread when attack occurs at the 2-position. The same argument can be applied to furan, but not to thiophen since additional Kekulé structures can be drawn and in this case the preferential orientation at the 2-position is less clearly defined. The substituents already present in the ring will affect the orientation but not sufficiently to prevent attack occurring at the 2- or 5-position if one of these is vacant.

always   S a donor or an   S an acceptor or
                      electron repeller   an electron attractor
                      (i.e. *ortho*/*para*   (i.e. *meta* directing)
                      directing)

Position of attack by an electrophile on furan, pyrrole, or thiophen carrying a single substituent S

We saw in Chapter 20 of Part 2 that the two available positions in naphthalene are not identical; electrophilic addition-with-elimination reactions result in substitution at the 1-position preferentially. A somewhat similar situation applies for the benzo-derivatives of pyrrole and thiophen. Both of these compounds undergo electrophilic addition-with-elimination reactions which lead to substitution in the 3-position. This would be analogous to attack at the 1-position in naphthalene. The preferential attack at the 3-position in indole and in thionaphthen is sometimes 'explained' by drawing attention to the fact that four of the canonical structures for the Wheland intermediates for attack at this position contain an intact benzene ring compared with two for the 2-position. We have pointed out previously that there is no theoretical justification for this type of argument and benzofuran, which must necessarily have similar canonical forms, undergoes attack preferentially at the 2-position. Nonetheless benzofuran is undoubtedly the odd man out in this series. We have repeatedly argued that in furan electron delocalization round the ring is much less established than in pyrrole and thiophen. Thus we could consider benzofuran as an *o*-alkoxystyrene, in which

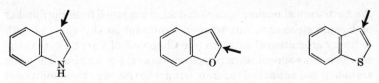

Site of electrophilic attack in indole, benzofuran, and thionaphthen

case we should expect electrophiles to attack the 2-position, the ω-position in the styrene molecule. A better explanation of the attack at the 3-position in indole and thionaphthen is that the positive charge is centred on the more electronegative heteroatom in two of the canonical forms of the intermediate whereas in attack at the 2-position only one canonical form has the positive charge on the heteroatom.

We have emphasized that pyrrole is attacked very readily even by quite weak electrophiles. Looking at this reactivity from another point of view we could describe pyrrole as a nucleophilic substance and we might expect pyrrole to add to suitable carbonyl double bonds and possibly to olefinic double bonds if they were sufficiently surrounded by electron-withdrawing or electron-accepting groups. The addition of pyrrole to maleic anhydride described on page 95 is an example of such an addition, and pyrrole does form condensation products quite readily with carbonyl compounds.

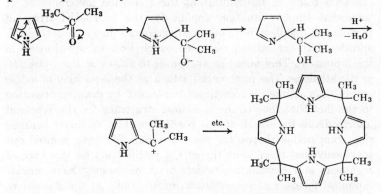

With formaldehyde, pyrrole forms a resin analogous to the phenol–formaldehyde resins used in the plastics industry (Part 1, Chapter 17, p. 171). Furan and thiophen undergo similar condensation reactions, although appreciably less readily than pyrrole. Indoles undergo these reactions and the compounds formed with aldehydes are notable for their similarity to the triphenylmethane dyestuffs (see Part 2, Chapter 19).

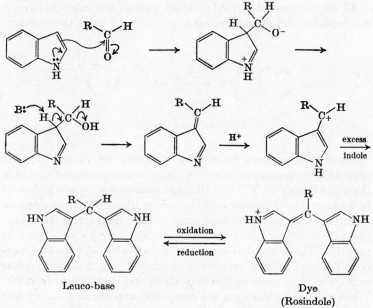

Leuco-base | Dye (Rosindole)

Indole is in fact sufficiently nucleophilic to behave as a nucleophile in a displacement reaction with an alkyl halide. We illustrate the principal compounds obtained by treating 2-methylindole with methyl iodide. In practice the reaction sequence is very complicated.

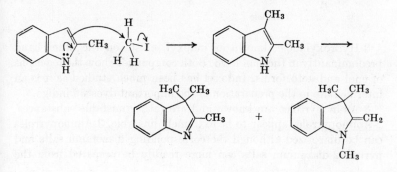

Pyrrole and indole also react with ethyl nitrate and ethyl nitrite in alkaline media to yield nitro-compounds and nitroso-compounds identical with those obtained by acidic nitration or nitrosation.

All the reactions described in which pyrrole and indole behave as nucleophiles used to be regarded as evidence for imine tautomerism.

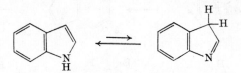

Hypothetical imine tautomerism

There is little evidence for such tautomerism and we should today regard these reactions as being analogous to the reactions of enamines (Part 2, Chapter 12, p. 148). Although tautomerism plays little part in the reactions of unsubstituted pyrrole or indole it is of considerable importance in the reactions of some of their derivatives. Hydroxy- and amino-derivatives of simple furans and pyrroles are extremely unstable. The benzo-derivatives are more stable and the hydroxy-derivatives of indoles have received much study. 2-Hydroxyindole exists almost entirely in this keto form, and is known as oxindole. When written in this form the compound appears as a lactam and indeed the ring can be reversibly opened and closed by treatment with alkali or acid.

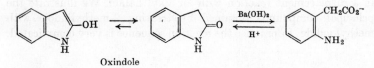

Oxindole

3-Hydroxyindole, known as indoxyl, exists on the other hand predominantly in the enol form. Both compounds show the reactions of enol and keto forms. Indoxyl has been much studied as it is an intermediate in the preparation of the important dyestuff indigo.

3-Aminopyrroles are known and are rather unstable substances. 2-Aminopyrroles appear to be extremely unstable. 3-Aminopyrroles can be diazotized although the corresponding diazonium salts and pyrrole-2-diazonium salts can more readily be prepared from the

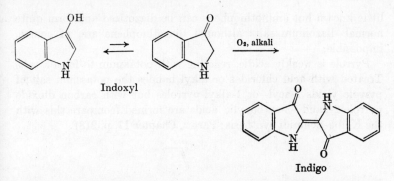

Indoxyl

O₂, alkali

Indigo

corresponding nitroso-compounds (see Part 2, Chapter 14, pp. 163—4). The resulting diazonium salts are interesting compounds. They are very stable and will not couple with phenols in neutral or acidic solutions. In alkaline solution they are converted into diazo-compounds analogous to diazocylcopentadiene.

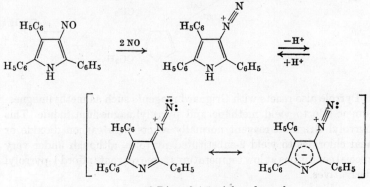

3-Diazo-2,4,5-triphenylpyrrole
(brown needles, m.p. 158°)

Since pyrroles can be nitrosated directly with nitrous acid it is possible to convert some pyrrole derivatives (such as the 2,4,5-triphenylpyrrole depicted) directly into the diazo-compound by treatment with an excess of nitrous acid, in a reaction exactly analogous to the formation of a diazonium salt from phenol (see Part 2, Chapter 15, p. 179). There is a very close relation between a phenolic diazonium salt and a diazo-oxide on the one hand and a pyrrole diazonium salt and a diazopyrrole on the other. Aminofurans are very unstable and little known but aminothiophens can be diazotized and form quite normal diazonium salts although diazothiophens are, of course, impossible.

Pyrrole is weakly acidic, reacting with potassium to form a salt. Treated with acid chlorides or alkyl halides the potassium salt of pyrrole yields 1-acyl- or 1-alkyl-pyrroles but with carbon dioxide pyrrole-2- and -3-carboxylic acids are formed (compare this with the Kolbe–Schmidt synthesis; Part 2, Chapter 17, p. 218).

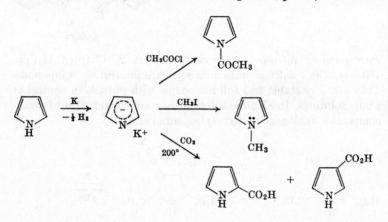

Pyrrole also reacts with Grignard reagents such as methylmagnesium iodide to yield methane and pyrrolylmagnesium iodide. This pyrrolyl Grignard reagent normally reacts with carbon dioxide or acid chlorides to yield 2-substituted pyrroles, although under very special conditions at low temperatures some reagents afford 1-pyrrolyl derivatives.

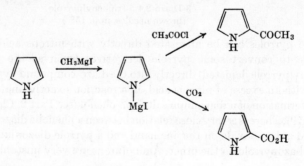

We have already discussed the reactions of benzo-derivatives of furan, pyrrole, and thiophen in which the benzene ring is fused to the 2- and 3-positions of the heterocyclic nucleus. There remain the benzo-derivatives in which the benzene ring is fused to the 3- and 4-positions, known as isobenzofuran, isoindole, and isothionaphthen (better called 2-benzofuran etc.); in the case of pyrrole there is also a compound in which the six-membered ring is fused on the 1- and 2-positions known as indolizine.

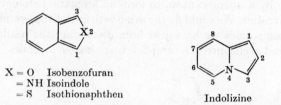

X = O   Isobenzofuran
= NH Isoindole
= S   Isothionaphthen

Indolizine

Isoindole itself is unknown, although *N*-substituted derivatives have been prepared. The characteristic reaction of isobenzofuran, isoindole, and isothionaphthen derivatives is participation in cycloaddition, which we illustrate by their reactions with maleic anhydride and transient dehydrobenzene (benzyne).

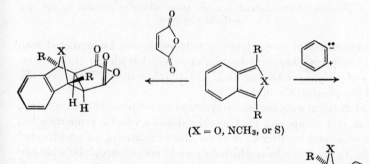

(X = O, NCH₃, or S)

Indolizine extremely readily undergoes electrophilic addition-with-elimination reactions characteristic of an aromatic nucleus, substitution occurring at the 1- and the 3-position.

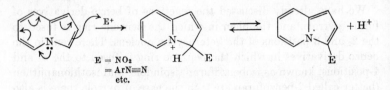

$$E = NO_2$$
$$= ArN \equiv N$$
etc.

The Azoles

We saw that we could conceptionally replace a CH group in benzene by a nitrogen atom to form an aromatic heterocyclic base called pyridine. We could do the same with any of the five-membered cyclic compounds we have just been discussing; the resulting five-membered cyclic nitrogen compounds are known as azoles.

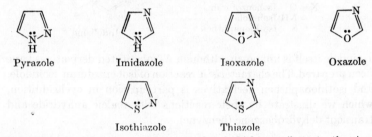

Pyrazole   Imidazole   Isoxazole   Oxazole

Isothiazole   Thiazole

The lone pairs of electrons shown are those which contribute to the ring π-electron system

The chemistry of these classes of compounds can be predicted with quite surprising accuracy by combining our knowledge of the chemistry of pyridine on the one hand with the chemistry of furan, pyrrole, and thiophen on the other.

Let us begin with the least important azoles, oxazole and isoxazole. Furan is a compound that weakly shows aromatic properties but whose general reactions are those characteristic of an unsaturated diene. Pyridine is a base which only undergoes electrophilic addition-with-elimination reactions with great difficulty. Oxazole and isoxazole are bases, they are considerably more stable in acid than furan but are decomposed by strong acids. The nitrogen atom makes electrophilic addition-with-elimination reactions extremely difficult and few such reactions have been reported.

Pyrazole and imidazole are stable aromatic substances. Imidazole is a much stronger base ($K_b = 1.2 \times 10^{-7}$) than either pyridine

($K_b = 2.3 \times 10^{-9}$) or pyrazole ($K_b = 3 \times 10^{-12}$). In both compounds tautomerism is possible. We can illustrate this with 4-methyl-imidazole, which because it cannot be distinguished from 5-methyl-imidazole, used to be called 4(5)-methylimidazole. Treatment with dimethyl sulphate gives *N*-methyl derivatives in which tautomerism is no longer possible, and two distinct compounds are formed.

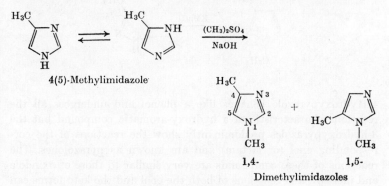

4(5)-Methylimidazole

1,4-   1,5-
Dimethylimidazoles

Both pyrazole and imidazole undergo electrophilic addition-with-elimination reactions, pyrazole invariably at the 4-position, while imidazole is attacked at the 2-position in neutral or alkaline media but in the 4-position in acidic media.

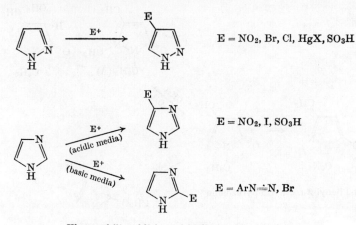

E = NO$_2$, Br, Cl, HgX, SO$_3$H

E = NO$_2$, I, SO$_3$H

E = ArN$=$N, Br

Electrophilic addition-with-elimination reactions

Both compounds are far more stable than pyrrole. Pyrazole is
particularly resistant to oxidation so that 1-phenyl-3-methylpyrazole
on treatment with permanganate yields 3-methylpyrazole (i.e., the
benzene ring is oxidized).

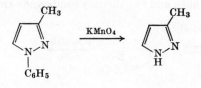

4-Hydroxypyrazole behaves like a phenol and undergoes all the
reactions characteristic of a hydroxy-aromatic compound but the
3-hydroxypyrazoles predominantly show the reactions of the corresponding enol (or lactam) and are known as pyrazolones. The
reactions of these compounds are very similar to those of oxindole
and indoxyl, and reactions of both the enol and the keto forms can
be observed depending on the reaction conditions. We illustrate
just the methylation of 1-phenyl-3-methyl-5-pyrazolone. Compound
(A) obtained by the reaction of methyl iodide and methanol is known
as antipyrine and once had considerable importance in medicine.

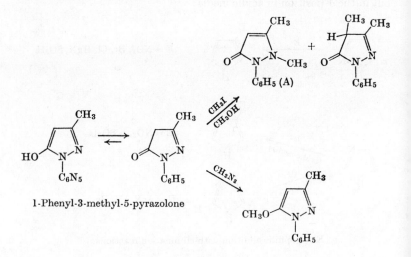

1-Phenyl-3-methyl-5-pyrazolone

5-Pyrazolones unsubstituted in the 4-position form very stable carbanions which take part in all the addition and substitution reactions we should expect.

The chemistry of imidazole and pyrazole shows the conflicting influences of the pyrrole-like and the pyridine-like character of the molecules. When we consider thiazole we must remember that thiophen is more benzene-like in its chemical properties than either furan or pyrrole. We should therefore expect thiazole and isothiazole to be much more like pyridine in their chemical properties than oxazole or imidazole. Thiazole is less basic than pyridine but it still forms quaternary salts with alkyl halides (usually but not exclusively on the nitrogen). As with pyridine or more particularly quinoline the quaternary hydroxide is in equilibrium with the carbinol base which in turn is in equilibrium with a thiol-aldehyde.

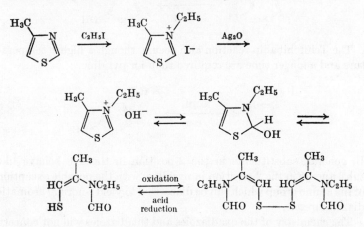

With electrophiles thiazole reacts in the same way as pyridine, i.e. to give a thiazolium salt, and just as pyridine can be sulphonated under very extreme conditions so thiazole can be sulphonated *and* nitrated using extreme conditions. Thiophen is more susceptible to electrophilic attack than benzene, and similarly thiazole (or more probably the thiazolium cation) is more susceptible to attack than pyridine (or more accurately the pyridinium ion).

Just as the resistance of pyridine to electrophilic addition-with-elimination reactions is contrasted by its susceptibility to nucleophilic attack, so thiazole undergoes nucleophilic addition-with-elimination

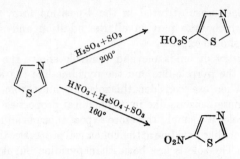

reactions. For example a halogen atom in the 2-position is very readily replaced.

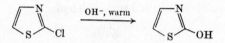

The Tchitchibabin reaction also occurs, though a higher temperature and a longer time are required than for pyridine.

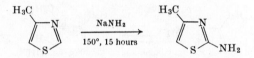

In general, substituents in the 2-position in thiazole behave like substituents in the 2-position in pyridine, with the notable exception of the amino group which can be diazotized to yield a normal aromatic diazonium salt.

The chemistry of the oxadiazoles and thiadiazoles will not concern us. A striking but not unexpected property of the triazoles and tetrazoles is the increasing acidity from pyrrole through imidazole to the triazoles and finally to tetrazole, which has an acidity approaching that of acetic acid. Electrophilic addition-with-elimination reactions are not known for 1,2,4-triazole or tetrazole but can occur with 1-methyl-1,2,3-triazole. Nucleophilic addition-with-elimination reactions are also not observed because the corresponding azole anion is formed. The triazoles and tetrazoles are very stable to oxidizing reagents and, in general, they show the properties of highly deactivated aromatic nuclei.

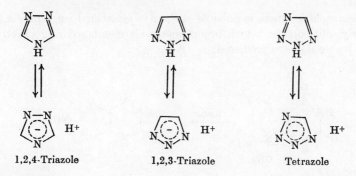

1,2,4-Triazole   1,2,3-Triazole   Tetrazole

The Purines

The purines, which consist of a pyrimidine ring fused on to an imidazole ring, are of the greatest biological importance.

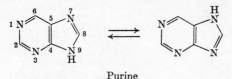

Purine

The azole proton is weakly acidic and treatment of purine with either dimethyl sulphate in alkali or with diazomethane in ether yields the 9-methyl derivative. Purine is a stronger base than pyrimidine but the site of quaternization depends on the conditions.

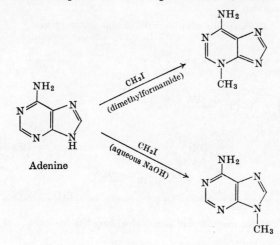

Electrophilic attack is possible in the five-membered ring and similarly, nucleophilic attack occurs in the six-membered ring exactly as we would have predicted.

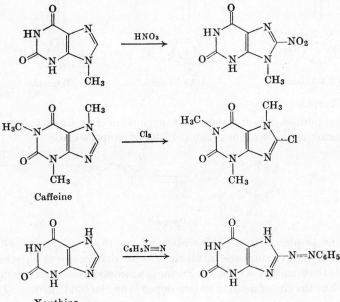

Caffeine

Xanthine

Electrophilic addition-with-elimination reactions

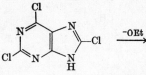

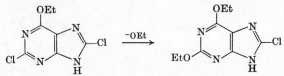

Nucleophilic addition-with-elimination reactions

Nucleophilic attack by the hydroxide ion can result in the displacement of alkoxy and amino groups as well as halogens.

The common stimulants of tea and coffee, namely caffeine and theobromine, are purines. Uric acid, the immediate cause of gout and the end nitrogen metabolic product in birds and reptiles, is also a purine. However, the most important purines are adenine and guanine which will receive much attention in Part 4.

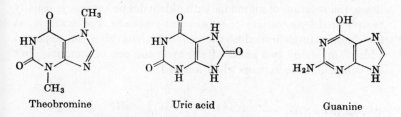

Theobromine   Uric acid   Guanine

The Synthesis of Heterocyclic Compounds

(a) *Nitrogen-containing heterocyclic molecules*

The most important reaction in forming nitrogen heterocyclic compounds is the reaction of an amino group with a carbonyl group.

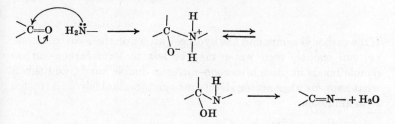

We have already described the reaction of α-diketones with *o*-phenylenediamine (benzene-1,2-diamine) to yield a quinoxaline derivative (Part 2, Chapter 18, p. 235).

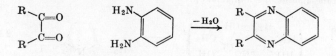

On p. 73 of the present volume we described reversible ring opening of 1-methylquinolinium hydroxide:

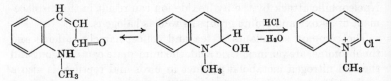

We saw in Part 1, Chapter 7, p. 72—73 that, whereas reaction of aldehydes or ketones with primary amines yields imines or Schiff bases, the reaction of ammonia with aldehydes or ketones is usually complicated by a variety of subsequent reactions. For example, at low temperatures benzaldehyde and ammonia give the expected adduct, involving two moles of aldehyde and one of ammonia, but the compound decomposes when warmed.

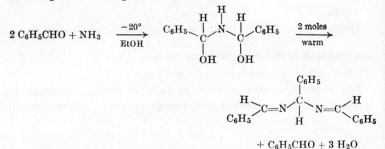

$$+ C_6H_5CHO + 3 H_2O$$

If the carbonyl compound has hydrogen atoms on the α-carbon atom, i.e. can enolize, then water may be lost to form carbon–carbon double bonds in place of carbon–nitrogen double bonds, and this is what happens when acetonylacetone or glutaconic aldehyde is treated with ammonia.

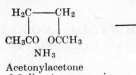

Acetonylacetone
(hexane-2,5-dione) + ammonia

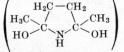

Hypothetical intermediate

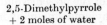

$$+ 2 H_2O$$

2,5-Dimethylpyrrole
+ 2 moles of water

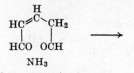

Glutaconic aldehyde
+ ammonia

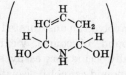

Hypothetical intermediate

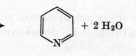

Pyridine + 2 moles of water

The preparation of 2,5-dimethylpyrrole by this reaction is of real synthetic value; but the preparation of pyridine from glutaconic aldehyde is not, since the easiest way of preparing glutaconic aldehyde is by ring opening of a pyridinium quaternary salt (see page 43). A similar reaction occurs with 1,3-diones and hydrazines to yield pyrazole derivatives (also phthalaldehyde and hydrazine to yield phthalazine).

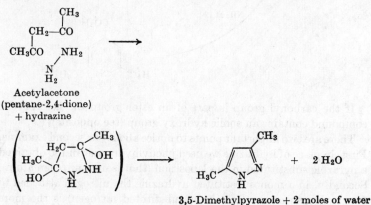

Acetylacetone
(pentane-2,4-dione)
+ hydrazine

3,5-Dimethylpyrazole + 2 moles of water

Imidazole derivatives can be prepared in the same way by using a 1,2-dione and an aldehyde with two moles of ammonia; but this involves four molecules coming together, and not surprisingly yields are not always as good as in the previous examples.

5

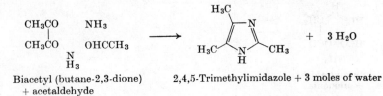

Biacetyl (butane-2,3-dione)   2,4,5-Trimethylimidazole + 3 moles of water
+ acetaldehyde
+ 2 moles of ammonia

The above preparation of pyridine involved an unsaturated aldehyde; a saturated 1,5-dicarbonyl compound, of course, yields initially dihydropyridine but oxidation usually occurs very easily, so much so that in some cases the dihydro-compound cannot be isolated.

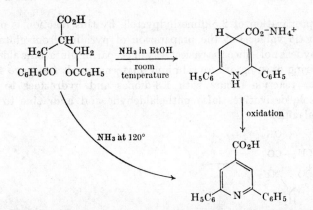

If the carbonyl group is part of an ester group the heterocyclic compound contains an enolic hydroxy group (see opposite).

There are two important points to notice about our second example. First, in place of hydrazine we used phenylhydrazine and so obtained a pyrazole substituted in the 1-position (that is on a nitrogen atom). Secondly, in a monosubstituted hydrazine the nitrogen attached to two hydrogen atoms (i.e. the unsubstituted nitrogen) is the more reactive and attacks the more reactive carbonyl group, i.e. the keto group rather than the ester group. We have drawn the intermediate hydrazone which can in fact be isolated.

In order to use this kind of reaction for the synthesis of the imidazole derivatives or pyrimidine derivatives we appear to require a

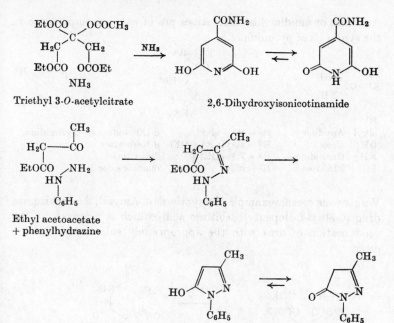

Triethyl 3-*O*-acetylcitrate

2,6-Dihydroxyisonicotinamide

Ethyl acetoacetate
+ phenylhydrazine

1-Phenyl-3-methyl-5-pyrazolone

1,1-diamino-compound. Such compounds are unknown and the nearest derivative is an amidine. Amidines can be formed from cyanides either by direct treatment with ammonium chloride, which gives the amidine hydrochloride, or via the imino-ether.

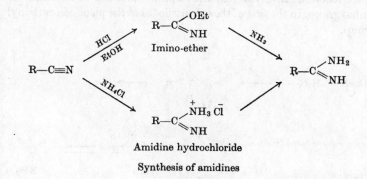

Imino-ether

Amidine hydrochloride

Synthesis of amidines

Amidines or amidine-like derivatives are of extreme importance in the synthesis of pyrimidines.

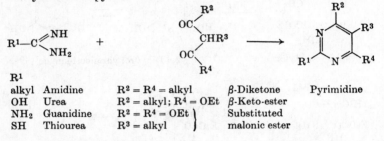

R^1		$R^2 = R^4 = $ alkyl	β-Diketone	Pyrimidine
alkyl	Amidine	$R^2 = R^4 = $ alkyl	β-Diketone	Pyrimidine
OH	Urea	$R^2 = $ alkyl; $R^4 = $ OEt	β-Keto-ester	
NH_2	Guanidine	$R^2 = R^4 = $ OEt ⎫	Substituted	
SH	Thiourea	$R^3 = $ alkyl ⎭	malonic ester	

We give one specific example, the synthesis of Amytal, the barbiturate drug (5-ethyl-5-isopentylbarbituric acid) which is prepared by the condensation of urea with the appropriately substituted malonic ester.

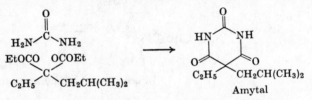

Nucleophilic addition of ammonia or an amino group to a carbonyl group is by far the most important route for the preparation of heterocyclic nitrogen-containing compounds. We should note, however, two other reactions involving the nucleophilic addition of an amine or amino group. In the first of these a nitrile takes the place of a carbonyl group.

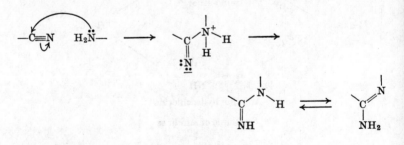

This is a reaction of quite general application but we illustrate it with just one example, the preparation of 4-aminouracil from **urea** and ethyl cyanoacetate.

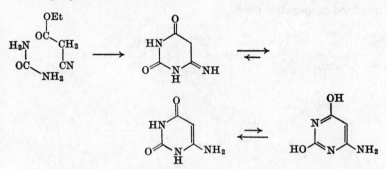

A related reaction is the preparation of imidazole derivatives from 2-amino-ketones and potassium thiocyanate.

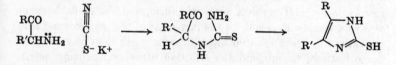

The thiol group can readily be removed, making this a synthesis of general utility. The second modification of the addition of a nucleophilic group to a carbonyl compound, we have to consider, is conjugate addition.

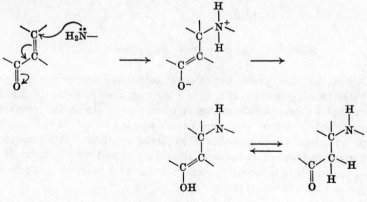

Aniline adds to acrolein or crotonaldehyde at the 3-position. The resulting adduct is set up to 'bite its own tail', and in the presence of acid a dihydroquinoline derivative is formed which may be readily oxidized to quinoline itself.

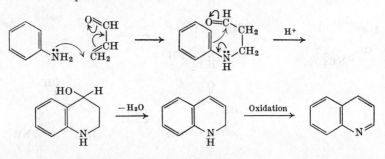

In practice, acrolein is unpleasant to handle and it can be formed *in situ* by the reaction of sulphuric acid on glycerol. The Skraup synthesis of quinoline consists of treating a mixture of aniline, nitrobenzene, and glycerol with sulphuric acid. The nitrobenzene acts as an oxidizing agent for the last stage of the reaction and is at the same time itself reduced to aniline which can then react further. Some authors recommend the addition of other oxidizing agents such as arsenic oxide. Surprisingly the yields from this reaction can be very high.

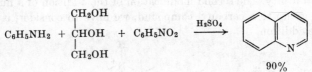

Skraup synthesis of quinoline

Notice that this synthesis involves an additional new principle. The imine-forming reaction is not itself the ring closure, as in the other examples considered above, but it is followed by the second reaction which leads to the ring formation. A simple example, also leading to quinoline, is the so-called Friedländer synthesis. This involves normal anil formation between acetaldehyde and *o*-aminobenzaldehyde, which is followed by condensation of a reactive methylene group with a carbonyl group.

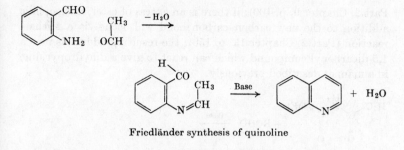

Friedländer synthesis of quinoline

Pyrroles can be synthesized in an analogous reaction known as the Knorr synthesis:

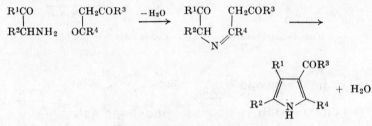

Knorr synthesis of pyrroles

In practice, α-amino-ketones are very unstable and are usually made *in situ* as depicted in the scheme.

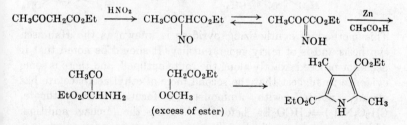

(excess of ester)

Two-stage reactions such as this are particularly important in the synthesis of pyridine, because 1,5-dicarbonyl compounds are required and these are usually prepared by aldol-type condensations performed in the presence of a base. Ethyl acetoacetate and an aldehyde condense in the presence of a base (the so-called Knoevenagel reaction;

Part 2, Chapter 9, p. 100); if there is an excess of ester, nucleophilic addition to the new carbon–carbon bond will occur via a Michael reaction (Part 2, Chapter 13, p. 156); the resulting adduct is now a 1,5-dicarbonyl compound which can react to give a dihydropyridine in a manner described previously.

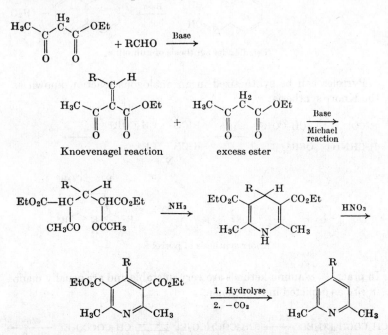

This method of synthesizing pyridine is known as the Hantzsch synthesis and is of fairly general utility. It should be noted that it need not proceed exactly along the route outlined, and there is some evidence to suggest that the second mole of ethyl acetoacetate has already reacted with ammonia to form β-aminocrotonate, $CH_3C(NH_2){=}CHCO_2Et$, before undergoing the Michael addition. Aldol or more specifically Knoevenagel-type condensations are involved in a wide variety of pyridine syntheses. Thus ethyl cyano-acetate condenses with acetylacetone in the presence of ammonia and the resulting 1,5-dicarbonyl compound reacts with ammonia to give a pyridine derivative (this reaction has been used in a synthesis of vitamin B_6).

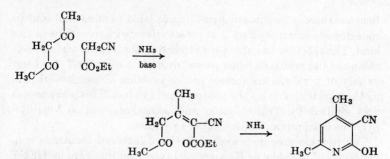

We saw in Chapter 9, p. 92, of Part 2 that not only would acetaldehyde condense with itself in the presence of a base to form aldol and crotonaldehyde, but that further polymerization was possible. If acetaldehyde and aqueous ammonia are heated in an autoclave at 200—250° for an hour a 60% yield of 5-ethyl-2-methylpyridine can be obtained together with 6% of 2-methylpyridine. We can rationalize this reaction in the following scheme.

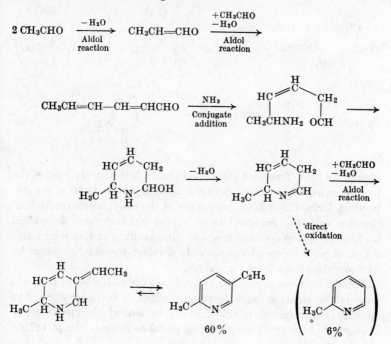

Because there is insufficient pyridine available from natural sources considerable research effort is at present devoted to reactions of this kind. The reaction has also been studied in the vapour phase, acetaldehyde and ammonia being passed over alumina at 300°. A greater variety of pyridines are formed in this reaction although 5-ethyl-2-methylpyridine remains the predominant product. This reaction was first studied by Tchitchibabin and sometimes 5-ethyl-2-methylpyridine is known as Tchitchibabin's pyridine.

The ring closures discussed so far have involved carbanion condensation of the aldol or Knoevenagel type, and these are by far the most important in the synthesis of nitrogen heterocyclic compounds. An alternative ring closure could involve a displacement reaction between a carbanion and a halogeno-compound. We illustrate two examples, the reaction of chloroacetone with ethyl acetoacetate and ammonia to yield ethyl 2,5-dimethylpyrrole-3-carboxylate and between the same ketone and acetamidine to yield 2,5-dimethylimidazole.

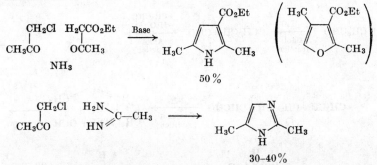

Neither of these reactions gives good yields. Notice in the case of the pyrrole synthesis we have drawn a furan in brackets (this is a by-product formed by the enolate anion of the ethyl acetoacetate condensing with the carbonyl bond of the chloro-ketone instead of undergoing a displacement reaction; the resulting adduct then forms a new enolate anion which can only displace the chlorine atom by nucleophilic attack from the oxygen).

Finally ring closure can involve an electrophilic addition-with-elimination reaction on a benzene nucleus. The Skraup reaction (p. 122) is one example of this kind of reaction, which is particularly important in the synthesis of the isoquinoline nucleus (see p. 127).

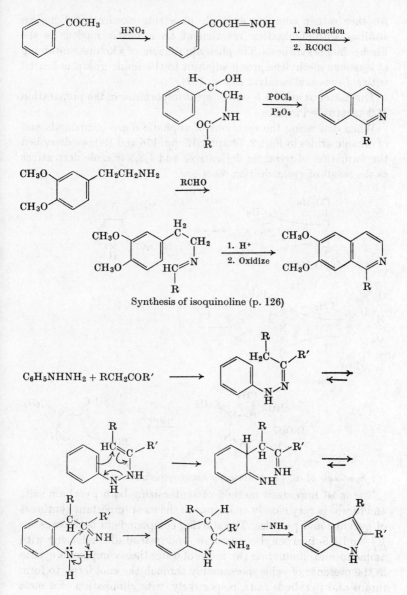

Synthesis of isoquinoline (p. 126)

Fischer indole synthesis (p. 128)

Another rather complex example of a ring closure involving an addition-with-elimination reaction at an aromatic nucleus is the Fischer indole synthesis. The phenylhydrazone of a ketone containing at least one methylene group adjacent to the imine group is heated with a Lewis acid catalyst (see p. 127).

This latter synthesis is one of great importance in the preparation of 2-substituted indoles.

When discussing the reactions of aliphatic diazo-compounds and of organic azides in Part 2, Chapter 15, pp. 178 and 190, we described the formation of pyrazole derivatives and 1,2,3-triazole derivatives as the result of cycloaddition reactions.

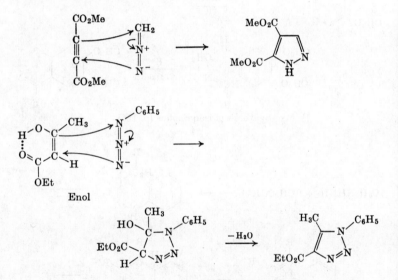

(b) *Synthesis of oxygen-containing heterocyclic molecules*

The most important method of synthesizing both pyrylium salts and furans is very closely analogous to the most important syntheses of pyridine and pyrroles. The starting compounds in both cases are 1,4- and 1,5-dicarbonyl compounds, and instead of condensing with ammonia and eliminating two moles of water these compounds cyclize in the presence of acids, presumably through the enol form, to form furans and pyrylium salts, respectively, with elimination of a mole of water in each case.

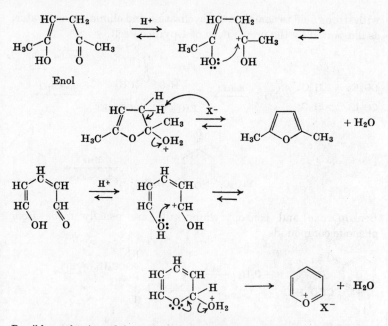

Possible mechanism of the acid-catalysed cyclisation of 1,4- and 1,5-diones to form furans and pyrylium salts

(Intramolecular proton transfer steps omitted)

Often an acid-catalysed condensation or addition can lead directly to the final cyclization in one reaction process.

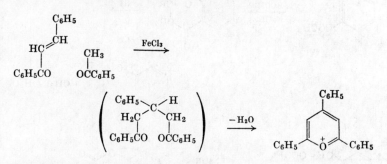

Alternatively, the 1,5-dicarbonyl compound can be formed from a usual series of base-catalysed carbanion reactions and then treated

with strong acid to complete the cyclization and elimination of water, as illustrated by the preparation of 4-pyrone itself.

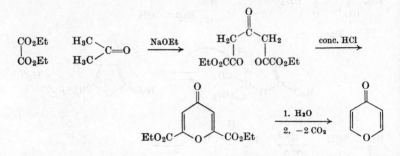

Benzopyrone and benzopyrylium salts are usually made from phenolic compounds.

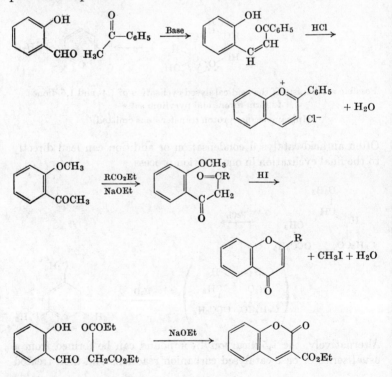

Each of these reactions involves the addition of a carbanion to form a side-chain to the benzene ring containing a carbonyl group in the γ-position, and in each case also the final ring closure involves condensation between the phenolic hydroxy group and the γ-carbonyl group. The synthesis of furans from 1,4-dicarbonyl compounds can be applied to the synthesis of oxazoles.

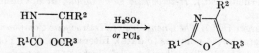

(c) *The syntheses of the thiophen derivatives*

Thiophen can be prepared from 1,4-dicarbonyl compounds by treatment with phosphorus trisulphide, usually at high temperatures in a sealed tube.

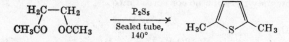

An alternative preparation is the base-catalysed condensation of 1,2-dicarbonyl compounds with diethyl thiodiacetate.

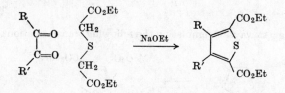

The diethyl thiodiacetate is readily prepared from sodium sulphide and ethyl bromoacetate ($2BrCH_2CO_2Et + Na_2S$).

Bibliography

R. M. Acheson, *An Introduction to the Chemistry of Heterocyclic Compounds*, 2nd Edn., Interscience, New York, 1968. (An excellent conventional text-book.)

A. R. Katritzky and J. M. Lagowski, *Principles of Heterocyclic Chemistry*, Methuen, London, 1967. (Full of original ideas but very concentrated.)

A. Albert, *Heterocyclic Chemistry: An Introduction*, Athlone Press, London, 2nd Edn., 1968. (Presents the subject in a very individual way.)

Other text-books include:

G. M. Badger, *The Chemistry of Heterocyclic Compounds*, Academic Press, New York and London, 1961.

M. H. Palmer, *The Structure and Reactions of Heterocyclic Compounds*, Edward Arnold, London, 1967.

Standard reference books include:

R. C. Elderfield (Ed.), *Heterocyclic Compounds*, in 6 Volumes, Wiley, New York.

A. Weissberger (Ed.), *The Chemistry of Heterocyclic Compounds*, a series of monographs (24 volumes to date), Interscience, New York.

E. H. Rodd (Ed.), *The Chemistry of Carbon Compounds*, Vol. IVA, IVB, and IVC, Elsevier, Amsterdam.

A. R. Katritzky (Ed.), *Advances in Heterocyclic Chemistry*, review articles published annually.

Problems

1. How may pyridine be converted into

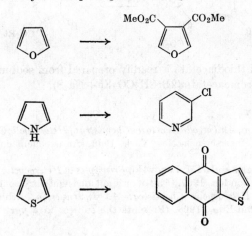

2. Suggest ways of completing the following transformation:

   ⟶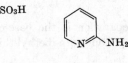

3. Starting from ethyl acetoacetate how would you prepare:

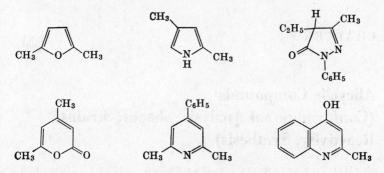

4. Suggest a synthesis of 2,6-diamino-4-hydroxypyrimidine, and indicate how the following may be prepared from it:

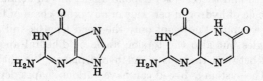

CHAPTER 3

Alicyclic Compounds
(Conformational Analysis, Shapes, Strains, Reactivity, Synthesis)

The shape of a molecule such as water is completely defined by the bond length and the H—O—H bond angle, but to define the shape of a molecule of hydrogen peroxide or any other chain of four atoms, a–b–c–d, we must know not only the bond lengths and a–b–c and b–c–d angles but also the torsion angle or dihedral angle, that is the angle between the a–b–c and b–c–d planes. We shall call this angle θ. The system a–b–c–d can have various shapes depending on the value of θ. These shapes are called conformations.

Conformational analysis is the study of the conformations of molecules and the interpretation of their chemical and physical properties on this basis. The concept of conformations and their representation as 'sawhorse' and Newman projections was introduced in Part 2, pp. 303—305. These projections are used to represent three-dimensional molecules on paper. Another way of doing this is to use thick lines for bonds projecting forward from the plane of the paper and broken lines for bonds projecting behind the paper. The three-dimensional shapes of molecules are very clearly illustrated by wire or plastic models which have the great advantage over drawings that their shape can change in the same way as that of the actual molecule. The frequent use of such models by the student will enormously facilitate his understanding of this and later chapters.

Acyclic Systems: Torsional and Steric Strain and Electrostatic Effects

Ethane can exist in the eclipsed form where $\theta = 0°$ or the staggered form where $\theta = 60°$ or in any of the conformations intermediate between these.

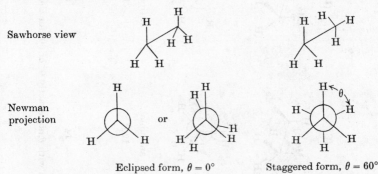

Sawhorse view

Newman projection   or

Eclipsed form, $\theta = 0°$ Staggered form, $\theta = 60°$

If the hydrogens of one methyl group were to rotate round the C—C bond through 360°, the molecule would pass through three eclipsed and three staggered conformations. Physical evidence (see p. 142) has shown that this rotation is not quite free and that in ethane a 120° rotation from one staggered conformation to another requires the temporary absorption of about 12 kilojoules per mole (kJ mole^{-1}) or 3 kcal mole^{-1}. The energy of the molecule changes with the dihedral angle roughly as shown

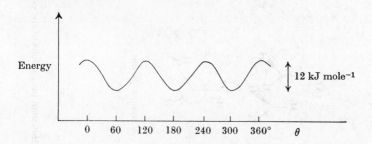

This energy difference between the staggered form and the eclipsed form is said to be due to torsional (or Pitzer) strain in the eclipsed form. The cause of this strain is by no means certain. It may be attributed to a repulsion between the electron pairs of the C—H bonds.

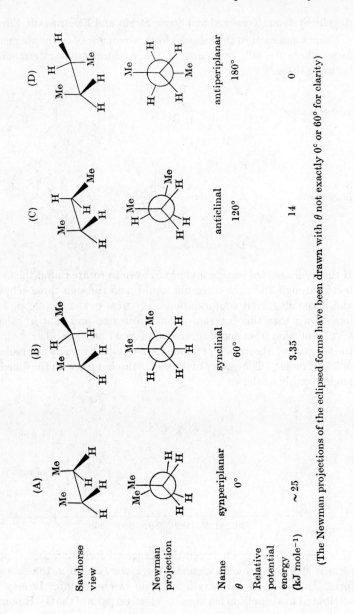

	(A)	(B)	(C)	(D)
Sawhorse view				
Newman projection				
Name	synperiplanar	synclinal	anticlinal	antiperiplanar
θ	0°	60°	120°	180°
Relative potential energy (kJ mole^{-1})	~25	3.35	14	0

(The Newman projections of the eclipsed forms have been drawn with θ not exactly 0° or 60° for clarity)

It is not due to steric repulsion between hydrogen atoms since these have van der Waals radii (see below) of 1.2 Å and are 2.3 Å apart in the eclipsed form and so they barely touch.

The energy barrier is sufficiently large to ensure that nearly all ethane molecules at room temperature are in a staggered conformation, but sufficiently small that at room temperature they change easily and frequently from one staggered conformation to another.

Butane could adopt any of a number of conformations (considering only rotation about the central C–C bond) including the two different eclipsed forms (A) and (C) and the two different staggered forms (B) and (D) shown. These conformations in which the dihedral angles between the C–Me bonds are 0°, 60°, 120°, and 180° are called the synperiplanar (A), synclinal (B), anticlinal (C), and antiperiplanar (D) conformations (see Part 2, p. 305).

The eclipsed conformations are subject to torsional strain, as is eclipsed ethane, but their energy is further increased by repulsions between non-bonded groups. There are usually attractive forces (van der Waals forces) between two atoms or groups of atoms which are not bonded to one another covalently or ionically. This attraction increases as the groups come closer, but at very short distances repulsive forces, increasing rapidly as the distance decreases, become dominant. The van der Waals radius of a group is defined so that the

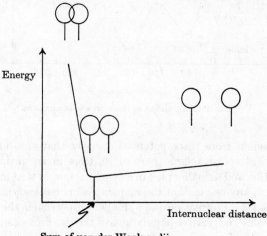

sum of the radii of two groups equals the distance between the central atoms when the energy is at a minimum. Thus the carbon nuclei of two non-bonded methyl groups cannot come closer than 4 Å without causing an increase in energy. The van der Waals radius of a methyl group is therefore 2 Å. Systems whose energy is raised owing to such non-bonded repulsions are said to be subject to steric or van der Waals strain (see p. 137).

In conformer (A) the carbon nuclei of the synperiplanar methyl groups are 2.6 Å apart. There is therefore considerable steric strain. Similarly, in the anticlinal form (C) the eclipsed methyl and hydrogen are 2.4 Å apart but the sum of their radii = 2 + 1.2 = 3.2 Å, so again there is some steric strain. Similar sums show that there should be less strain in the synclinal and virtually none in the antiperiplanar conformations. The actual potential energy differences between the conformations found by experiment are given on p. 136. The energy of the molecule changes with the dihedral angle roughly as shown.

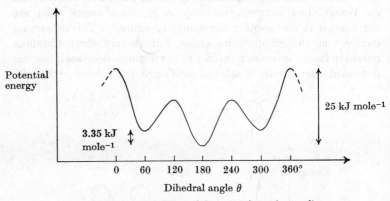

(The detailed shape of the curve is conjectural)

It is obvious from these potential energies that any one butane molecule will spend a large part of its time in an antiperiplanar conformation and a smaller part in a synclinal one, or that in a sample of butane at any one instant the majority of molecules will be in the antiperiplanar conformation and a smaller proportion in the synclinal conformation. The relative populations of the two forms can be found if we know the free energy difference between them. The difference in potential energy is equivalent to a difference in enthalpy, ΔH^0, between the conformers. To calculate the free energy difference, $\Delta G^0 = \Delta H^0 - T\Delta S^0$, we must therefore establish the difference in entropy between them. Conformers can have different entropies because of differences in solvation etc., but also because there are different numbers of equivalent but non-superimposable versions of each conformer. Thus there are two forms of synclinal butane which are mirror images, but only one form of the antiperiplanar conformer.

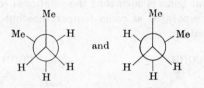

The two synclinal conformers of butane

The entropy of the synclinal form is therefore greater. Since there are two forms in one case and one in the other, the entropy difference is $R\ln 2$. In the change from synclinal butane to antiperiplanar butane at 27°C (300°K), both the enthalpy and the entropy decrease. $\Delta H^0 = -3.35$ kJ mole^{-1} and $\Delta S^0 = -R\ln 2$ Joules degree^{-1} mole^{-1}.

$$\Delta G^0 = \Delta H^0 - T\Delta S^0$$
$$= -3.35 - \frac{300(-R\ln 2)}{1000}$$
$$= -3.35 + 1.73$$
$$= -1.62 \text{ kJ mole}^{-1}$$

since $R = 8.314$ J degree^{-1} mole^{-1}.

We can now calculate the equilibrium constant, K, for this conversion:

$$\log K = -\frac{\Delta G^0}{2.3RT} = \frac{1.62 \times 1000}{2.3 \times 8.31 \times 300} = 0.28$$

Therefore $K = 1.9$.

This represents the ratio of the two populations. Therefore at 27°C and at equilibrium, butane is about 65% in the antiperiplanar form and 35% in the synclinal form. At lower temperatures the equilibrium constant will be greater.

The conformers of butane are easily interconvertible. The conformationally isomeric forms of a molecule can only be separated at room temperature if the activation energy barrier for interconversion is at least 80 kJ mole^{-1} ($\sim$20 kcal mole^{-1}). The barrier to interconversion of the conformers of butane is much less than 80 kJ mole^{-1} so that it is not possible to prepare at room temperature a sample of butane in which all the molecules are in one conformation.

In certain biphenyls such as the one shown, the barrier to rotation about the central bond is such that the rotational (conformational) isomers *can* be separated at room temperature but they are interconverted on heating.

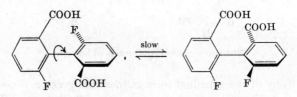

In this case it happens that the two conformations are enantiomeric. Therefore the substance is resolvable at room temperature but racemizes on heating.

Butane, and tetramethylene chains in larger molecules, prefer to adopt the antiperiplanar conformation. The preferred conformation of any straight-chain alkane would be expected to be a regular planar zig-zag which has an antiperiplanar conformation about each C–C bond. This arrangement is in fact found in crystalline paraffins and crystalline polyethylene.

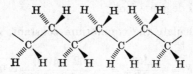

The preferred conformation of many molecules or parts of molecules containing other functional groups can usually be correctly predicted by assuming that hydrogen atoms in a CH_2 group can be replaced with no essential stereochemical change by a lone pair of electrons on an oxygen, nitrogen, or sulphur atom and that double bonds can be considered as two bent single bonds. Thus the preferred conformation of ethers, esters, ketones, amines, etc., is found to be as shown.

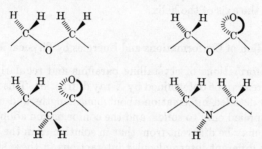

van der Waals repulsion and repulsion of bonding pairs are not the only forces affecting conformations. Electrostatic repulsions or attractions and hydrogen-bonding may be important. Thus in the case of the propyl halides the synclinal conformation is preferred by 2 kJ mole^{-1} (0.5 kcal mole^{-1}) over the antiperiplanar. This may be due to favourable dipolar attractions overcoming the expected steric repulsion, and in ethylene chlorohydrin the synclinal form appears to be more stable than the antiperiplanar by about 4 kJ mole^{-1} (1 kcal mole^{-1}) perhaps owing to hydrogen-bonding.

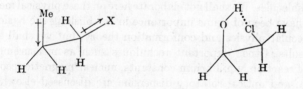

Hydrogen-bonding is also a major factor in determining the conformation of protein chains (Part 1, p. 160).

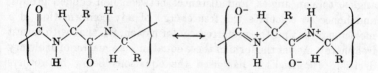

Rotation about the CO–N bonds is hindered because of the electron delocalization (Part 1, p. 90). In protein fibres the chain takes up the form of a very broad helix with about 3.6 amide units per turn, which is held in place by $C{=}O \cdots HN$ hydrogen bonds all roughly parallel to the axis of the helix.

Determination of Conformations and Energies by Physical Methods

The conformations of crystalline paraffins and regularly ordered protein fibres can be determined by X-ray diffraction. This technique gives very detailed information about bond lengths and angles but it can be applied only to solids, and the conformation adopted in the solid state may be different from that in solution or in the gas phase because of different intermolecular interactions in these states. For molecules such as ethane, butane, and cyclohexane, which are volatile, valuable information can be obtained about conformations in the vapour from electron diffraction studies and from microwave absorption spectra which are due to changes in rotational energy levels. Vibrational spectra (infrared and Raman) can also give useful structural information. The measurement of thermodynamic data (heats of combustion, heats of hydrogenation, and heat capacities) allows calculation of molecular enthalpies and entropies, and the comparison of these quantities with spectroscopically determined values has provided much information on the internal motions of small molecules. We shall not elaborate here on these physical methods which have been of prime importance in establishing the basic facts of molecular physics and conformation theory, but we shall accept the results. Other important techniques such as measurement of rates of reaction, equilibrium constants, nuclear magnetic resonance spectra, and optical rotatory dispersion are discussed elsewhere in this chapter.

Sufficient basic data are now available to allow us to set up equations relating the energy of a group of atoms to the bond lengths, bond angles, torsion angles, and distances between non-bonded groups, and hence to calculate the free energy of any conformation of a molecule or to find the preferred conformation, i.e. that with lowest free energy. As yet the form of these equations is a matter of arbitrary decision. Hendrickson has used the equations below but others (Wiberg, Allinger) have used different ones. The dependence of van der Waals strain on the nature and distance of the atoms is particularly difficult to estimate.

Angle strain (kJ mole^{-1}) = $k_a(\Delta\tau)^2$ for each angle

$\Delta\tau$ is the deviation in degrees of the C–C–C angle from the tetrahedral value, and k_a is a constant estimated from spectral data to be 0.08 kJ mole^{-1} degree^{-2}.

Torsional strain (kJ mole^{-1}) = $k_t(1 + \cos 3\theta)$ for each C–C bond

θ is the torsion angle in degrees, and k_t is a constant estimated to be 5.6 kJ mole^{-1}.

van der Waals strain (kJ mole^{-1}) = $10^{(4.62-2r)} - 205r^{-6}$

for each pair of interacting non-bonded hydrogens

r is the distance in Å between the hydrogen nuclei.

In calculating free energies, the effects of bond stretching, symmetry, partition functions, and zero-point energy may also be included. The calculation of the free energy for any set of atomic coordinates and the adjustment of the coordinates to minimize the total free energy can be done by computer. By this method estimates have been made of the relative energies of different conformations of molecules such as butane, substituted cyclohexanes, and medium sized cycloalkanes, and calculation of the probable preferred conformation by this method is much quicker than determination of it by X-ray or electron diffraction.

Cycloalkanes: Angle Strain

We must now ask what happens if we join up the ends of a chain to make a ring. If we consider cyclobutane, it is obvious that a new factor appears. Whereas in butane and n-alkanes the C–C–C angle is about 112°, in cyclobutane it cannot be more than 90° (and in fact it is less; see p. 179). Deformation of a C–C–C angle by ±5° from its normal tetrahedral value of 109½° causes an increase in energy of about 2 kJ mole^{-1} (0.5 kcal mole^{-1}) and by ±10° of 8 kJ mole^{-1} (2 kcal mole^{-1}) for each such angle in the molecule. Since cyclic systems may of necessity have abnormal angles, they may be subject to another form of strain, angle strain. The amount of angle strain, like torsional and steric strain and electrostatic interactions, will depend on the conformation. The preferred conformation will be that compromise where the total free energy is least.

(a) *Cyclohexane*

The most important and the most studied ring system is cyclohexane. If cyclohexane were flat the bond angles of 120° would cause considerable angle strain. If we make a model (try it) of cyclohexane deliberately with normal tetrahedral carbon atoms (angle $109\frac{1}{2}°$) we find that there are two distinct *types* of conformations possible *both free of angle strain*, a chair form, which is also free of torsional strain and of steric strain, and a flexible (or twist or skew) form. The chair is rather rigid, but the flexible form can easily change into other flexible forms as a result of rotations about the C–C bonds. In so doing it passes through a symmetrical boat form. In the boat form there are eclipsed interactions along the sides of the boat and steric interactions between the internal hydrogens on the bow and stern. In the skew form these interactions are reduced but not eliminated.

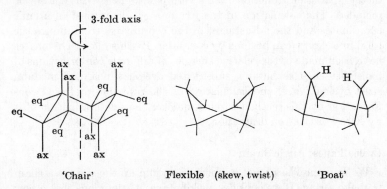

'Chair'   Flexible   (skew, twist)   'Boat'

Calculation and experimental evidence agree that the skew form is less stable than the chair by about 22 kJ mole^{-1}, and the boat form about 6.7 kJ mole^{-1} less stable still. After allowance for entropy differences, this accords with there being no more than 0.2% of cyclohexane in the flexible form at room temperature. Most substituted cyclohexanes also exist almost entirely in the chair form. Exceptions exist owing to special effects.

Let us look at our model of this ubiquitous chair again. There are six axial C–H bonds marked 'ax' all parallel to one another and to the three-fold axis of the model. The six equatorial bonds marked 'eq' are approximately in the plane of the ring and each is parallel to the equatorial bond at the opposite side of the ring and to two ring bonds. All six carbon atoms have identical environments. All six equatorial hydrogens are identical and all six axial hydrogens are identical. The completely staggered arrangement means that the *cis*-substituents A and C (ax,eq) and B and D (eq,ax) are as far away from one another as the *trans*-substituents B and C (eq,eq) (i.e. $\theta = 60°$), while the other two *trans*-substituents A and D (ax,ax) are antiperiplanar ($\theta = 180°$).

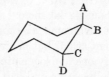

It is probable that the actual shape of cyclohexane chairs is a little different from this model in that the axial bonds are a little splayed outwards away from the three-fold axis and from one another, which means that the ring is a little flattened giving a C–C–C angle of 111.5°. This relieves the slight steric interaction between the 1,3-diaxial hydrogens at the expense of increasing the torsional strain (1,2-interactions). Thus the actual conformation is a compromise between those factors.

Reduction of the dihedral angle between two 1,2-*cis*-substituents of a cyclohexane chair, for example by *cis*-fusion of a small ring, flattens the ring. This requires relatively little energy (torsional strain is increased but 1,3-diaxial interactions are reduced). Reduction of the dihedral angle between two 1,2-*trans*-substituents puckers the ring, increasing torsional *and* 1,3-diaxial interactions. This requires a considerable increase in energy.

There are of course two chair forms, identical in cyclohexane but not in substituted derivatives.

The chair forms are fairly rigid but they can be interconverted via a flexible form. The process is called ring inversion and results in all equatorial substituents becoming axial and all axial ones equatorial.

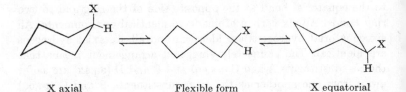

X axial   Flexible form   X equatorial

The exact geometrical changes which occur in this process are not
known but the energy barrier, $\Delta G^{\ddagger}$, to ring inversion can be found
by experiment to be about 42 kJ mole^{-1} (10 kcal mole^{-1}). This is
appreciable but can be overcome at room temperature, so that
interconversion is frequent at 25° but has virtually ceased at −100°.
At room temperature any one hydrogen atom of cyclohexane spends
half of its time in an axial position and half in an equatorial position
and is changing frequently from the one environment to the other,
whereas at −100° hydrogens will remain in axial or equatorial positions
for a long while.

This can be seen from the nuclear magnetic resonance spectrum
of cyclohexane. At room temperature cyclohexane stays in any one
chair form for about 10^{-5} second. The NMR spectrometer 'observes'
the proton for about 10^{-2} second during which period it has changed
environment many times. All twelve protons therefore appear to
the spectrometer to have identical averaged environments and hence
have the same chemical shift. The spectrum at room temperature
therefore consists of a single sharp line. At −100° ring inversion is
slow. The protons remain in an axial or equatorial position during
the time of observation. Two bands are observed in the spectrum
at different chemical shifts corresponding to protons in two distinct
environments.

Monosubstituted cyclohexanes would also be expected to exist
almost entirely in chair forms but now the two chair forms are
different with the substituent axial in one and equatorial in the other
(see p. 147).

The barrier to the interconversion of the two chair forms is still about
42 kJ mole^{-1} so they are rapidly interconverting at room temperature.
But they need not be equally stable. When the substituent X is a
carbon atom, the 1,3 X–H distance in the axial form is 2.55 Å whereas
in the equatorial form the 1,2 X–H distance is 2.8 Å. If X is large,
van der Waals steric compression is therefore less when X is equatorial.

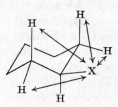

Axial form
X—H distance 2.55 Å

Equatorial form
X—H distances shown all 2.8 Å

On this simple argument the conformer with the substituent equatorial should be more stable, the difference in energy between the two forms being dependent on the size of X, on the length of the C–X bond, and possibly on other steric or polar factors not yet considered.

In methylcyclohexane, the 1,3- and 1,5-interactions in the axial form are actually synclinal CH_3,CH_2 interactions. We know that in butane a synclinal interaction of this type raises the energy by 3.35 kJ mole^{-1} (0.8 kcal mole^{-1}) compared with the antiperiplanar form. In the equatorial form of methylcyclohexane there are no synclinal carbon–carbon interactions but in the axial form there are two. The axial form of methylcyclohexane might therefore be expected

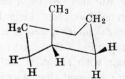

Axial form of
methylcyclohexane

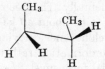

Synclinal form
of butane

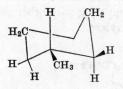

Equatorial form of
methylcyclohexane

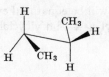

Antiperiplanar
form of butane

to be about 6.7 kJ mole^{-1} (1.6 kcal mole^{-1}) less stable than the
equatorial one. Experimental determinations of the difference give
values of about 7.1 kJ mole^{-1} (1.7 kcal mole^{-1}) in the liquid phase.
There is almost no entropy difference between the axial and equatorial
forms of methylcyclohexane. Therefore for the interconversion
ax → eq, $\Delta G^0 = -7.1$ kJ mole^{-1}. The equilibrium constant is given
by $RT \ln K = -\Delta G^0$. Therefore,

$$\log K = \frac{-\Delta G^0}{2.3\,RT} = \frac{7.1 \times 1000}{2.3 \times 8.31 \times 300} = 1.24$$

and
$$K \approx 20$$

At room temperature (300°K) the ratio of the populations of the two
conformers is thus twenty to one, or 5% of methylcyclohexane is in
the axial form and 95% in the equatorial form.

The energy differences between the axial and equatorial conformations of several monosubstituted cyclohexanes have been measured.
As can be seen from the table, the equatorial preference varies with
the size of the group, its shape, and its distance from the ring. Thus,
although bromine atoms are larger than chlorine atoms, the C–Br
bond length is greater. The value for Bu^t has not been measured
experimentally. The free energy difference has been calculated (see
p. 143) to be about 40 kJ mole^{-1} (10 kcal mole^{-1}) and it is known
that cyclohexanes carrying a Bu^t group will, if necessary, adopt a
twist conformation rather than a chair with an axial Bu^t group.
The axial form must therefore be more strained than a twist form
and so at least 20 kJ mole^{-1} higher in free energy than the equatorial
form. The energy difference is such that in t-butylcyclohexane less
than 0.01% of the molecules are in the axial form, and the t-butyl
group will be equatorial even when competing with other groups as
in ethyl *cis*-4-t-butylcyclohexanecarboxylate.

Free energy differences between axial and equatorial conformations of substituted cyclohexanes $C_6H_{11}X$.

X =	Me	Et	Pr^i	Bu^t	Ph	CN	CO_2H	CO_2Et	OH	Cl	Br
$-\Delta G$ for ax→eq (kJ mole^{-1})	7.1	7.5	9.2	>20	12.5	0.8	5.4	5.0	2.9	2.1	2.1
(kcal mole^{-1})	1.7	1.8	2.2	>4.5	3.0	0.2	1.3	1.2	0.7	0.5	0.5

(b) *Polycyclic systems*

Let us look at some more complex systems containing cyclohexane rings. Polycyclic systems in general can be divided into four groups: spiro systems such as spiro[5,5]undecane, bridged systems such as bicyclo[3,2,1]octane, fused systems such as bicyclo[4,3,0]nonane, and catenanes in which the rings have no atoms in common but are joined merely by being linked through one another like the links of a chain.

Spiro[5,5]undecane   Bicyclo[3,2,1]octane   Bicyclo[4,3,0]nonane

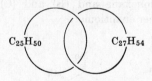

A catenane

The figures in square brackets are the lengths of the chains joining the bridgeheads. The skeletons are numbered from one bridgehead to the other via the longest bridge and back as shown above. Should there be more bridges, their position in the original skeleton is indicated by superscripts as in tricyclo[2,2,1,0^{2,6}]heptane.

6

based on

Tricyclo[2,2,1,0^{2,6}]heptane   Bicyclo[2,2,1]heptane

For the moment we will restrict our attention to fused systems based on cyclohexane rings. Fusion of two cyclohexane rings gives the compound bicyclo[4,4,0]decane or decahydronaphthalene, usually called decalin.

Decalin

(It is conventional to use a numbering system based on that of naphthalene)

In fact, two isomeric decalins exist in which the hydrogen atoms at the ring junction are respectively *cis* and *trans* to one another.

In order to represent the stereochemistry in two-dimensional formulae it is conventional to draw the bonds to substituents which are on the near side of the ring as thick lines and bonds to substituents behind the ring as broken lines. This also defines the absolute stereochemistry. Thus (A) represents one of the enantiomers of *cis*-2-chloro-1-methylcyclohexane and (B) and (C) are both ways of representing the other enantiomer.

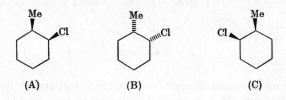

(A)   (B)   (C)

In polycyclic systems it is convenient to indicate the configuration of the hydrogen atoms using either thick or broken lines as before or using large dots or no dots as shown for the two decalins.

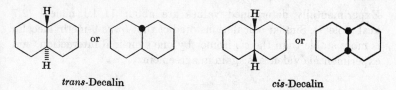

trans-Decalin   *cis*-Decalin

If you make models you will find that for *trans*-decalin only one chair-chair conformation is possible while for *cis*-decalin two chair-chair conformations of equal energy are possible. In the *cis* case, inverting both rings changes the molecule from one chair-chair form to the other. The conformation of all *trans*-decalins is therefore fixed (unless boat forms develop) but substituted *cis*-decalins have the choice of two conformations and can change from one to the other easily.

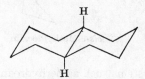

trans-Decalin

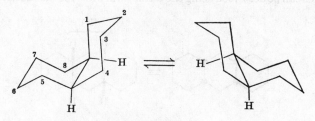

cis-Decalin

In *cis*-decalin there are three synclinal interactions between methylene groups in different rings. Using the conventional numbering shown these are between positions 1 and 7, 3 and 5, and 1 and 5. There are no such interactions in *trans*-decalin. If we assume again that each synclinal interaction raises the energy of the molecule by about 3.35 kJ mole^{-1} (0.8 kcal mole^{-1}), we conclude that the *trans*-isomer should be more stable by about 10 kJ mole^{-1} (2.4 kcal mole^{-1}).

Experimentally determined values are about 11 kJ mole⁻¹ (2.7 kcal mole⁻¹). Similar calculations predict that *trans*-9-methyldecalin is more stable than the *cis*-isomer by one synclinal interaction, and experimental evidence is again in agreement.

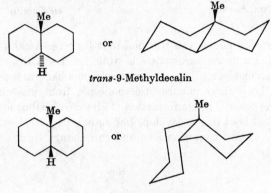

trans-9-Methyldecalin

cis-9-Methyldecalin

For polycyclic systems such as perhydroanthracene or perhydrophenanthrene the relative configurations of the various asymmetric centres can be described using the terms *cis* or *trans* for the substituents

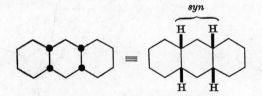

cis-syn-cis-Perhydroanthracene

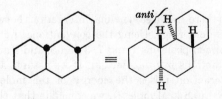

trans-anti-trans-Perhydrophenanthrene

on carbon atoms joined by a ring-fusion bond, and *syn* and *anti* for substituents on carbon atoms not joined by a ring-fusion bond. The geometrical relationships are then listed starting from one end as shown at the bottom of page 152.

One group of polycyclic compounds that has been particularly studied is the steroids. This group possesses a tetracyclic skeleton in which the rings A to D are fused *trans-anti-trans-anti-trans* as shown and this forces the molecule to adopt a conformation which is rigid and immobile (except for ring A which is free to become a boat). This rigidity, and the fact that compounds of this type are known with a great range of functional groups in a variety of positions, makes steroids a fascinating testing ground for our ideas on conformations. The presence of the axial angular methyl groups results in the back or bottom face of the molecule as drawn being more accessible to reagents.

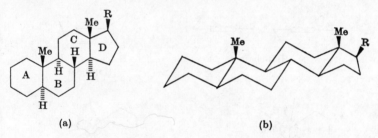

(a)   (b)

The steroid ring system, showing (a) its absolute configuration and the relative configuration of the seven asymmetric centres, and (b) its absolute and relative configuration and its conformation

(c) *Cyclohexanones*

The introduction of one sp^2 centre into cyclohexane as in cyclohexanone and methylenecyclohexane results in one C–C–C angle being increased from 110° to 120°. The rings are still chairs but flatter, and the barrier to ring inversion is less.

The removal of one axial substituent reduces 1,3-interactions but the equatorial bonds on C-2 and C-6 are now almost coplanar with the carbonyl group so that 1,2-interactions may be important, particularly if the equatorial groups on C-2 are large. If these equatorial groups are also polar there may be electrostatic interactions with the carbonyl group.

(d) *Cyclohexenes*

The introduction of an endocyclic double bond has a marked effect on the conformation of a ring. In cyclohexene four of the ring atoms (and two hydrogen atoms) are necessarily coplanar. Of the conformations available the two equivalent half-chair forms are about 11 kJ mole^{-1} (2.7 kcal mole^{-1}) more stable than the half-boat.

Half-chair   H   H

                                                            Half-boat

In the half-chair forms the dihedral angles between substituents on the two types of single bond are not 60° but are as shown.

If two of these substituents are joined to form a fused ring a further complication arises. Consider the *trans*-octahydronaphthalenes (A) to (D). In (A) the 50° dihedral angle imposed on the saturated ring at the ring fusion by the cyclohexene ring will cause a flattening of the cyclohexane ring which requires little energy (see p. 145). In (B), the 70° dihedral angle at the ring fusion puckers the saturated ring and brings the axial substituents closer together. Isomer (A) is therefore the more stable of the two. This difference in stability will be increased if axial groups are present in the saturated ring. Thus

(C) is more stable than (D), and the steroid (E) is more stable than its isomer (F).

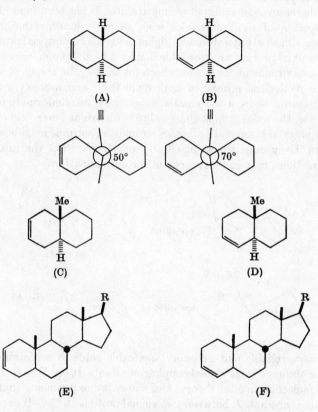

(A)

(B)

(C)

(D)

(E)

(F)

Effects of Conformation on Physical and Chemical Properties of Cyclohexanes

Let us now look at some of the physical and chemical properties of substituted cyclohexanes in the light of our knowledge of conformations and relative stabilities.

(a) *Nuclear magnetic resonance spectra*

The NMR spectrum of cyclohexane at low temperatures shows separate signals for the axial and equatorial protons (see p. 146).

Substituents in axial and equatorial positions in any cyclohexane which is prevented from inverting have different environments and may therefore show different chemical shifts. It has been found from the spectra of rigid molecules such as *trans*-decalins that axial protons almost always absorb at higher fields than equatorial protons, i.e. the axial protons are more shielded by 0.1—0.7 p.p.m. depending on the environment. It has also been found that the chemical shifts of the methylene groups of hydroxymethyl- and acetoxymethyl-cyclohexanes show a systematic dependence on conformation. In this case, the axial groups almost always absorb at lower fields than the equatorial isomers. In all cases strongly anisotropic neighbouring groups, i.e. groups whose physical properties are not the same in all directions, may cause reversal of the normal pattern.

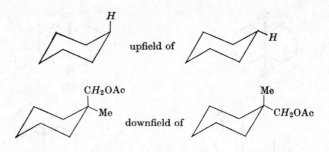

A more reliable and generally applicable guide to conformations can be obtained from vicinal coupling constants. It has been deduced from molecular-orbital theory, and shown by experiment, that the coupling constant, J, between two vicinal protons H–C–C–H depends on the torsion angle, θ, between the C–H bonds. Very roughly $J = k \cos^2 \theta$. The constant k has a value of about 10 for $0° < \theta < 90°$, and 16 for $90° < \theta < 180°$, but its value depends on both the C–C–C angles in C–$\overset{|}{C}$H–$\overset{|}{C}$H–C and on the nature of the substituents attached to all of those carbon atoms. It is common to draw a graph of J against θ but the shape of this curve depends on the compounds studied. For cyclohexanes with no highly electronegative substituents, J is about 10 Hz when $\theta = 0°$, about 0 Hz when $\theta = 90°$, and about 16 Hz when $\theta = 180°$. The introduction of electronegative substituents reduces the magnitude of all vicinal coupling constants.

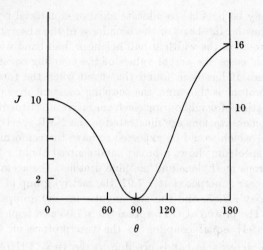

In a cyclohexane there are three types of vicinal coupling constant.

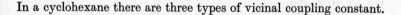

$J_{trans}(\text{ax,ax})$ $\theta = 180°$. J should be about 16 Hz

$J_{trans}(\text{eq,eq})$ $\theta = 60°$. J should be about 3 Hz

$J_{cis}(\text{eq,ax})$ $\theta = 60°$. J should be about 3 Hz

We can see that an axial proton in an unsymmetrical cyclohexane will be coupled to the two different axial protons at C-2 and C-6 with large coupling constants, and to the two different equatorial protons at C-2 and C-6 with small coupling constants. It will therefore appear as a broad multiplet. An equatorial proton will be coupled with small coupling constants to its four neighbours and should appear as a fairly sharp multiplet.

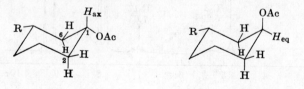

Thus it may be possible to allocate axial or equatorial positions to such protons on the basis of the broadness of the absorption band, usually quoted as the width at half height or 'half band width'.

In simple cases the actual values of the coupling constants may be obtained. It has been found that even when the torsion angle between protons is the same, the coupling constant may depend on the orientation of neighbouring electronegative substituents.

These generalizations are illustrated in the NMR spectrum of the steroid (A) which would be expected to have the conformation (B). Its NMR spectrum shows a broad band centred about τ 8.4 due to the numerous methylene and methine protons. The acetate methyl groups appear superposed at τ 7.97, the methyl group of the acetyl side-chain at τ 7.83, and the two angular methyl groups at τ 9.08 and 9.30. The proton at C-17 appears at τ 7.03 as a triplet owing to approximately equal coupling to the two protons on C-16. The other two identifiable bands are due to the two $\rangle$CHOAc groups.

We see from the conformation of the steroid that one acetate is axial at C-12 and the other equatorial at C-3, i.e. the C-12 proton is equatorial and the C-3 proton axial. We would therefore expect the C-12 proton to appear at lower field than the C-3 proton. This assignment is confirmed by the line widths. The upfield proton is a very broad multiplet as we expect for an axial proton coupled to four different vicinal protons with two large and two small coupling constants. The equatorial proton at C-12 is coupled to only two protons, at C-11, by small eq-ax and eq-eq constants and is a much narrower band.

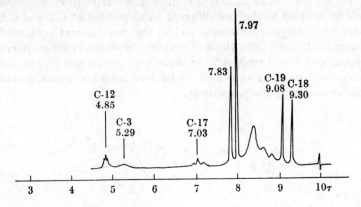

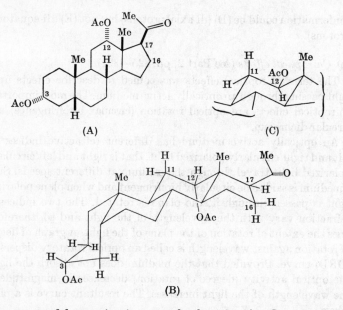

(A)

(C)

(B)

A more careful examination reveals that, in (A), $J_{12eq-11eq}$ and $J_{12eq-11ax}$ are both 2.5 Hz. The torsion angles in both cases are 60°. However in epimeric steroids with equatorial acetoxy groups at C-12 as in (C), $J_{12ax-11ax}$ ($\theta = 180°$) is 11 Hz and $J_{12ax-11eq}$ ($\theta = 60°$) is 5 Hz. Thus the diaxial coupling constant is larger than the eq-ax or eq-eq constants but the values of J_{eq-ax} depend on the orientation of the acetoxy group.

Using information of this kind it is possible to determine the orientation of a substituent at a known site in a rigid cyclohexane of known conformation. Conversely, if the relative geometry is known, we may be able to establish the conformation. For example, if we observe that $J_{A, B}$ is 12 Hz for *trans*-2-methylcyclohexanol, the

(D)

(E)

conformation could be (D) (di-axial protons) but not (E) (di-equatorial protons).

(b) *Chiroptical effects* (see Part 2, pp. 325—331)

The term chiroptical effects was coined to describe effects upon light passing through an optically active medium. The most important chiroptical effects are optical rotation (circular birefringence) and circular dichroism.

An optically active medium has different refractive indices for left and right circularly polarized light, that is right and left circularly polarized light travel through the medium at different speeds. Such a medium is said to be circularly birefringent and when plane polarized light is passed through it, the plane is rotated. The two indices of refraction vary with the wavelength of the light and so, therefore, does the extent of rotation of the plane of the light. A graph of degree of rotation against wavelength is called an optical rotatory dispersion (ORD) curve. Provided that the medium does not absorb the light, the optical activity (degree of rotation) decreases in magnitude as the wavelength of the light increases. The resultant curve is a plain curve.

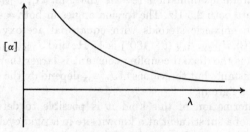

Plain positive ORD curve

Comparison of the rotatory powers of similar structures has been of great value in *relating* absolute configurations, but it is not easy to deduce the absolute configuration of a compound from the sign and magnitude of its optical rotation.

If a medium absorbs light in the wavelength range studied, the ORD curve is not plain but S-shaped and is said to be anomalous. If the right and left components of plane polarized light passing through an optically active medium are absorbed to different extents, the medium is said to exhibit circular dichroism and the light becomes elliptically polarized. The molecular ellipticity $[\theta] = 3300\,\Delta\epsilon$ where $\Delta\epsilon = \epsilon_L - \epsilon_R$ and ϵ_L and ϵ_R are the molecular extinction coefficients (Part 2, p. 356) for the left and right components of the light. The ellipticity varies with wavelength, and a graph of $[\theta]$ or $\Delta\epsilon$ against wavelength is called a circular dichroism (CD) curve.

A substance whose molecules are chiral will show optical rotation, and at wavelengths where it absorbs light it will also show circular dichroism. The combination of these phenomena is called the Cotton effect after the French physicist who studied it. The development between 1955 and 1960 of automatic equipment has permitted the

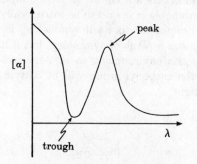

ORD curve showing a positive
Cotton effect superposed on a plain
positive curve

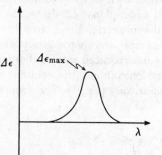

CD curve showing a positive
Cotton effect

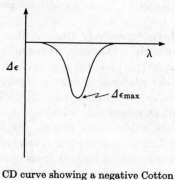

CD curve showing a negative Cotton
effect

rapid measuring and recording of ORD and CD curves. A measured ORD curve might show an S-shaped inflection in an otherwise plain curve. If on going from longer to shorter wavelengths the curve rises to a peak then falls to a trough, the Cotton effect is said to be positive. If the trough comes at longer wavelength the Cotton effect is negative. For a CD curve, if $\Delta\epsilon$ rises to a positive maximum the Cotton effect is positive, if it falls to a negative maximum the effect is negative.

A Cotton effect is associated with a chromophore (Part 2, p. 359). A molecule may contain several chromophores and give rise to a complex curve showing multiple Cotton effects. If the orbitals from and to which electron promotion occurs are chiral, as for example in a non-planar 1,3-diene, the chromophore is said to be intrinsically dissymmetric. More often the chromophore is itself symmetric, for example a carbonyl group which has a plane of symmetry, but if it is incorporated into a dissymmetric environment as in 3-methyl-cyclohexanone, the orbitals of the carbonyl group will be dissymmetrically perturbed to some extent.

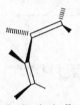

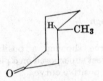

An intrinsically
dissymmetric chromophore

An intrinsically symmetric but
dissymmetrically perturbed chromophore

Ketones have been extensively studied since the Cotton effect occurs at a conveniently measurable wavelength (that of the UV maximum at about 300 nm) and the relation between the sign of the Cotton effect and the molecular goemetry is fairly well understood. However, many other dissymmetrically perturbed chromophores show Cotton effects at measurable wavelengths, for example, esters, lactones, disulphides, azides, nitro groups, xanthates, nitrites, and phthalimides. The electronic absorption of most alkenes and alcohols occurs at too short wavelength for easy measurement. The following discussion is restricted to cyclohexanones, but similar sector treatment can be applied to other chromophores.

Light of wavelength about 300 nm causes the promotion of one electron from a non-bonding $2p_y$ oxygen atomic orbital which is in the C–C–C plane, to an antibonding π molecular orbital the lobes of which lie in a plane perpendicular to the C–C–C plane ($n \rightarrow \pi^*$ transition) (Part 2, p. 357).

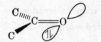

Filled $2p_y$ atomic orbital

Vacant π_{xz}^* molecular orbital

When this process occurs in the carbonyl group of a ketone in which the orbitals are dissymmetrically perturbed, the transition probabilities, and so the ϵ values, for left and right circular polarized light are different. The sign and magnitude of the Cotton effect depends on the disposition of the perturbing atoms relative to the carbonyl group and on their proximity to it, and they can be predicted on the basis of a semi-empirical *octant rule* founded on experimental observations and supported later by theoretical analyses.

The space round the carbonyl group of cyclohexanone can be divided into eight octants separated by three planes, the plane of symmetry of the whole molecule (which is the nodal plane of the oxygen $2p_y$ orbital), the plane in which the atoms $O{=}C\overset{\displaystyle C}{\underset{\displaystyle C}{\Big\langle}}$ lie, and a plane perpendicular to those two intersecting the $O{=}C$ bond. (These latter planes are nodal planes of the π_{xz}^* orbital.) If we now view this assembly along the direction $O \rightarrow C$, all the carbon and hydrogen atoms will lie in the four rear octants and will appear in projection as shown. (This is clearly seen from a wire model of the molecule.) The octant rule states that atoms lying in the different octants produce Cotton effects with the signs shown. Substituents close to the carbonyl group such as 2_{ax} have greater effects than substituents remote from it. The effects of carbon atoms 3 and 5 of the cyclohexanone cancel exactly. The C-4 methylene group, carbon atoms 2 and 6, and the substituents at positions 2_{eq} and 6_{eq} lie on nodal planes and do not contribute to the Cotton effect.

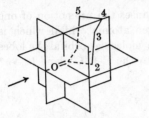

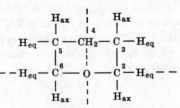

Planes separating the octants
of cyclohexanone

Projection of cyclohexanone as seen
when viewed along the O=C bond

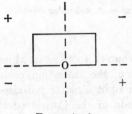

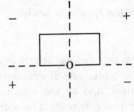

Rear octants

Front octants

Atoms falling in these octants produce Cotton effects with the signs shown

(+)-3-Methylcyclohexanone shows a positive Cotton effect. If we make only the assumption that the ring is a chair, this molecule could have any of the geometries (A)—(D).

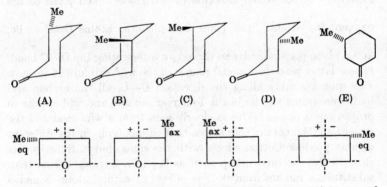

(A) and (B) are the possible conformations of one enantiomer, and (C) and (D) are conformations of the other. The corresponding octant projections (rear octants only are involved) show that only (A) and

(C) should show positive Cotton effects. If we assume that, whatever the configuration, the methyl group will be equatorial, then (+)-3-methylcyclohexanone must be (A). Alternatively, if we knew from other data that the (+)-enantiomer has the absolute configuration (E) we could deduce from the ORD or CD curve that it must adopt (predominantly) the conformation with the methyl group equatorial (again assuming it is a chair). The octant rule for cyclohexanones thus allows us to deduce the configuration if we know the conformation, or to deduce the probable conformation if the configuration is known. For example, *cis*-9-methyl-1-decalone of known configuration indicated shows a positive Cotton effect. Two chair-chair conformations (F) and (G) are possible. In (F), C-8 (decalin numbering) is axial to the cyclohexanone ring and α to the carbonyl group. It therefore would produce a strong negative Cotton effect. The rest of the molecule is further away and so less important. In (G) the major effect is due to the α-methyl group. The methylene groups at positions 6 and 7 lie very close to the plane separating front and rear octants, and so make little contribution. Conformation (G) should have a positive Cotton effect and so this is the predominant conformation of this molecule.

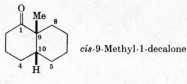

cis-9-Methyl-1-decalone

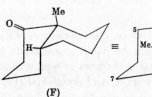

(F)

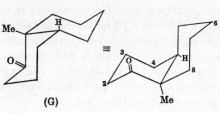

(G)

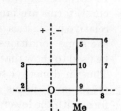

(c) *Infrared spectra*

The force constant for stretching the C–X bond of a substituted cyclohexane depends on the axial or equatorial orientation of the group X. Thus the C–OH stretching band of equatorial cyclohexanols usually occurs at higher wavenumbers than that of the epimeric axial alcohol.

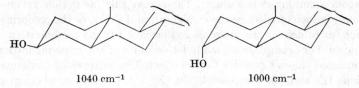

1040 cm⁻¹   1000 cm⁻¹

The frequency of the C=O stretch of cyclohexanones is sensitive to the presence of adjacent dipoles. Thus equatorial α-chloro- or α-bromo-cyclohexanones show carbonyl bands at higher wavenumbers than the unhalogenated ketones whereas the presence of axial α-halogens has little effect. The close alignment of the dipoles in the equatorial case has been held to be responsible for this effect (see p. 153).

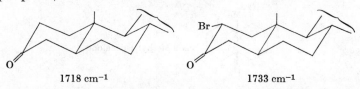

1718 cm⁻¹   1733 cm⁻¹

This dependence of the carbonyl frequency on the orientation of the bromine is shown by 2-bromocyclohexanone itself which exists in solution as a mixture of axial and equatorial forms in equilibrium. In carbon tetrachloride the axial (1730 cm⁻¹) and equatorial (1742 cm⁻¹) forms are present in the ratio 3:1, as can be shown from the relative intensities of the two bands observed. (Cyclohexanone under the same conditions absorbs at 1730 cm⁻¹.) The IR intensities can be used to show that the equilibrium constant depends on the solvent. Thus in dimethyl sulphoxide the axial (1729 cm⁻¹) and equatorial (1743 cm⁻¹) forms are present in almost equal amounts. The axial form with the lower dipole moment and lower dipole–dipole repulsion is therefore more favoured in the solvent of lower polarity.

Infrared spectroscopy also reveals intramolecular hydrogen-bonding which may be useful in determining conformations. Thus *trans*-cyclohexane-1,2-diol shows intramolecular bonding which is possible only if it adopts the eq-eq conformation.

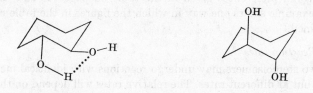

(d) *Equilibrium constants*

If two diastereoisomers are interconvertible, the more stable one should predominate at equilibrium. This difference in stability may be due to a difference in strain. Thus *cis*- and *trans*-1-decalones are interconvertible in base via the enolate ion, and the *trans*-isomer predominates owing to the greater stability of the *trans*-decalin system (see page 151).

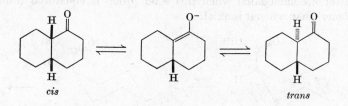

cis   *trans*

Similarly, the *cis*- and *trans*-isomers of ethyl 4-t-butylcyclohexane-carboxylate are interconvertible in the presence of sodium ethoxide. At 78° in ethanol, the equilibrium mixture contains 85% of the *trans*-isomer. The difference in free energy between the isomers is

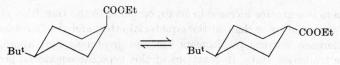

Ethyl *cis*- and *trans*-4-t-butylcyclohexanecarboxylate

therefore about 5 kJ mole^{-1} (1.2 kcal mole^{-1}). In the *cis*-isomer one of the substituents (actually the ester group for reasons given on page 148) must be axial while in the *trans*-isomer both substituents can be equatorial. The difference in stability is thus in line with the equatorial preferences given in the table on page 149 and indeed this example shows one way in which the figures in the table can be obtained.

(e) *Reaction rates*

Two stereoisomers may undergo reactions with identical mechanisms but at different rates. The relative rates will depend on the free energies of activation in the two cases, that is on the increase in free energy from the ground state to the transition state for the two isomers.

For instance, *cis*- and *trans*-ethyl 4-t-butylcyclohexanecarboxylates are both hydrolysed in base by the same mechanism but the *trans*-isomer reacts twenty times faster than the *cis*-isomer with sodium hydroxide in aqueous ethanol. In both cases there is an increase in bulk of the ester group in the slow step due to a change from planar to tetrahedral geometry and to increased solvation. This increase is better accommodated when this ester group is equatorial (*trans*-isomer) than when it is axial.

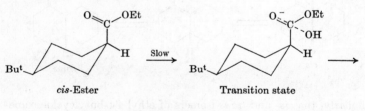

cis-Ester   Transition state

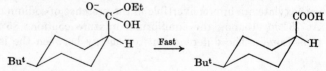

There is a greater increase in strain on going to the transition state in the axial ester than in the equatorial ester, and the free energy difference between the isomers becomes greater as they approach the transition state. In reactions of this type the equatorial group reacts faster than the axial. Thus equatorial hydroxycyclohexanes are more rapidly esterified and their esters more rapidly hydrolysed than axial isomers.

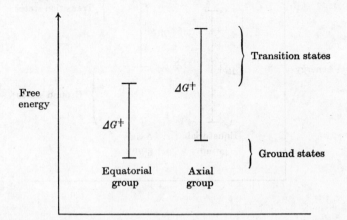

If, however, there is reduction in steric strain in the transition state for both isomers, the faster reaction should be that in which there is the greater relief of strain, namely the reaction of the more strained starting material (see diagram, p. 170). Thus *cis*-4-t-butylcyclohexyl tosylate reacts with acetic acid under S_N1 conditions four times faster than the *trans*-isomer.

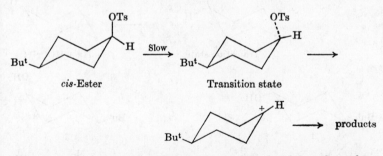

Note, however, that *trans*-2-methyl-*cis*-4-t-butylcyclohexyl tosylate solvolyses much more slowly than its equatorial epimer, *cis*-2-methyl-*trans*-4-t-butylcyclohexyl tosylate. Dissociation of axial cyclohexyl tosylates is probably assisted by participation by a *trans* β-hydrogen. The availability of such assistance may be more important than relief of steric strain.

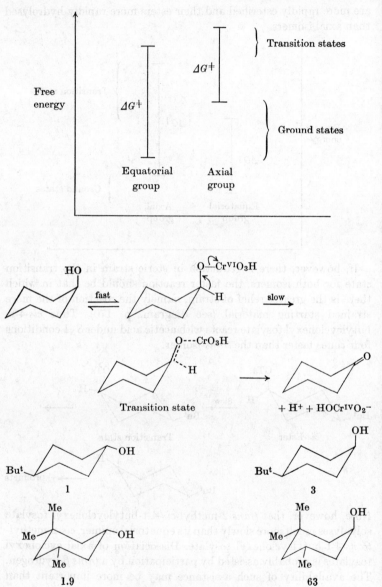

Relative rates of oxidation

A slightly more complex example of the same effect is provided by the oxidation of alcohols by chromic acid (see Part 1, p. 101). It is known that this reaction proceeds by initial fast and reversible formation of a chromate ester. This breaks down by a slow $E2$ type elimination of H^+ and $HOCrO_2^-$ to give the ketone. There is a decrease in steric strain in the transition state for this slow step with the result that axial alcohols are oxidized faster than equatorial ones, and hindered axial ones faster still (see bottom of p. 170).

Another case where the strain in the transition state affects reaction rates is that involving formation of a new fused ring. In the oxidation of cyclohexane-1,2-diols with lead tetraacetate or periodic acid, the *cis*-isomer reacts faster than the *trans* in spite of the fact that the two hydroxy groups are the same distance apart in both. The reactions involve formation of a five-membered cyclic ester, and it is easier to do this if the hydroxy groups are *cis* since the new ring pulls them together and flattens the cyclohexane ring (see p. 145). In the *trans*-isomer the five-membered ester ring would require increased puckering of the cyclohexane. Similar effects are observed in the formation of acetonides.

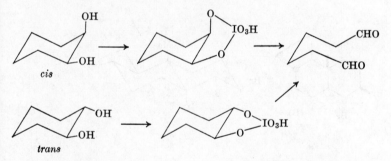

(f) *Reaction path*

A given reaction may require a certain geometry of the transition state which is possible for one isomer but impossible for another, because of their conformations. Thus $E2$ reactions proceed most easily if the groups involved are antiperiplanar (Part 2, pp. 338—344). The steroid (A) reacts much faster with iodide ions than its isomer (B) in which the bromine atoms can only become antiperiplanar if ring A becomes a boat. Note that both (A) and (B) are *trans*-dibromocyclohexanes.

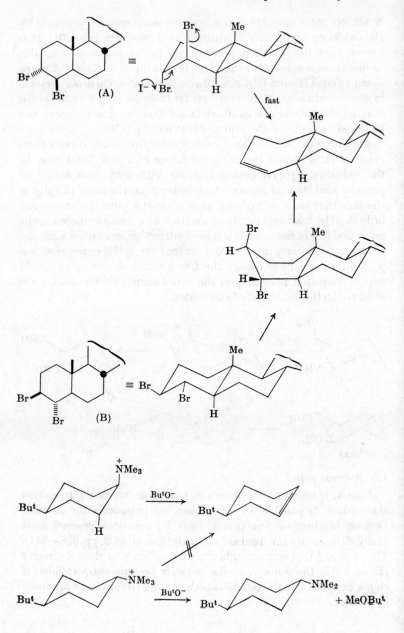

In the case of the *cis-* and *trans-*cyclohexylammonium salts shown, the *cis-*isomer can readily undergo an *E*2 elimination but the *trans-*isomer, where this is impossible (in the chair form anyway), undergoes an S_N2 reaction at one of the *N*-methyl groups instead (see bottom of p. 172).

Formation of epoxides from halohydrins also requires an anti-periplanar arrangement. The steroids (C) and (D), in which the bromine and hydroxy are axial, readily form epoxides on treatment with base. The isomer (E) does not.

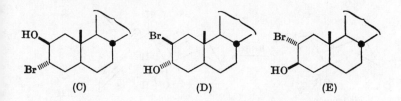

       (C)   (D)   (E)

Carbonium ion rearrangements usually require the incoming and outgoing groups to be antiperiplanar at the beginning of the reaction (cf. Part 2, p. 338). Treatment of *trans-*2-aminocyclohexanol (in which both groups are equatorial) with nitrous acid gives an unstable diazonium salt which decomposes with migration of the ring bond which is antiperiplanar to the departing nitrogen. The *cis-*isomer under the same conditions gives two products. This is because this compound may exist in two conformations of about equal stability. In one the ring contraction is still possible. In the other the only group antiperiplanar to the departing nitrogen is a hydrogen atom. Migration of it produces cyclohexanone (see top of p. 174).

In cyclohexane chairs no two adjacent substituents can be syn-periplanar. This conformation is required by *cis-*elimination reactions such as the pyrolysis of amine oxides which proceeds via a five-membered transition state (see Part 2, p. 344). In the amine oxide shown on p. 174, the only hydrogen atoms which can be synperiplanar to the nitrogen are those of the methyl group. The observed products demonstrate the importance of the conformational requirement of the transition state since the less stable olefin is formed almost exclusively.

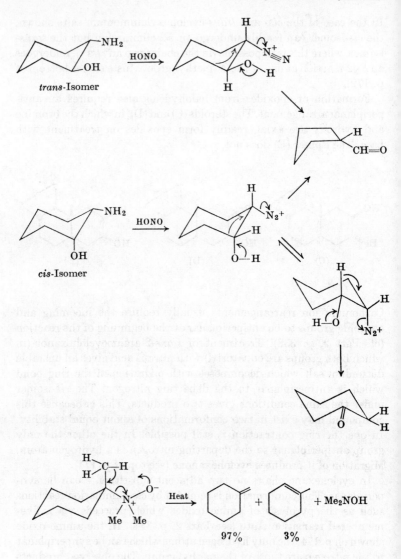

The pyrolysis of esters is also a *cis*-elimination but the ring formed in the transition state is this time six-membered and need not be planar. A synclinal arrangement of the hydrogen and the ester will-then be adequate. The ring hydrogens can therefore react, and the olefin with the endocyclic double bond predominates in the products.

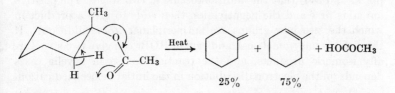

25%   75%

(g) *Stereospecific reactions*

One of the commonest reactions of ketones is their substitution in the α-position by the action of electrophiles on the enolate ion. Deuteration, bromination, alkylation, and aldol and Michael condensations are examples. For a cyclohexanone, the enolate ion could react with, say, Br⁺ in two ways. In both, the Br⁺ approaches perpendicularly to the plane of the double bond.

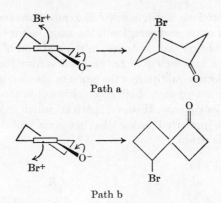

Path a

Path b

Path (a) will result in the formation of the product in a chair form with the bromine axial, while path (b) will result in the formation of a flexible form. Path (a) is preferred except in cases where the new axial substituent causes severe steric strain. The initial product of bromination, alkylation etc. of enols or enolate ions is that in which a chair ring is formed and in which the substituent is axial. The initial product may epimerize rapidly, however.

The same principle applies to the ionic addition of reagents such
as bromine or hypobromous acid to cyclohexenes. We know (Part 2,
pp. 77 and 344) that the addition occurs in two stages. The positive
ion adds first and the negative ion then adds to give a product in
which the two new groups are antiperiplanar (*trans*-addition). If
the olefin is unsymmetrical, addition of HOBr can give two structur-
ally isomeric products, and the reaction course in acyclic cases
depends on the electron distribution in the initial delocalized cation.

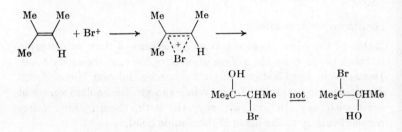

In addition reactions involving cyclohexenes, however, preferential
formation of a chair conformation is the main factor controlling the
direction of addition. For instance, in the addition of Br₂ or HOBr
to cyclohexenes, attachment of Br⁺ occurs first from the less hindered
side of the molecule. Addition of the anion to the intermediate cation
can then occur in two ways, both of which result in the new substituents being antiperiplanar. However, path (a) which produces a chair-
shaped ring is preferred unless the two axial substituents cause
excessive steric strain.

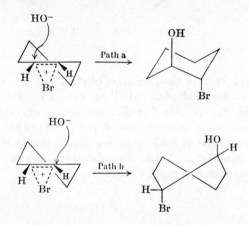

This preference for the formation of a chair may control the structure (i.e. the site of attachment of the hydroxy group) as well as the stereochemistry of the product, and in certain cases the product may be that structural isomer which would not be expected on the basis of inductive effects alone. Although the initial preferred product is that with the new substituents both axial on a chair ring, rapid epimerization of the initial adduct may sometimes occur.

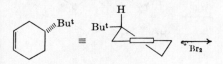

1-t-Butylcyclohex-3-ene

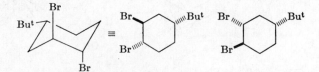

trans-3-*cis*-4-Dibromo-
1-t-butylcyclohexane

Not
formed

Control of product stereochemistry

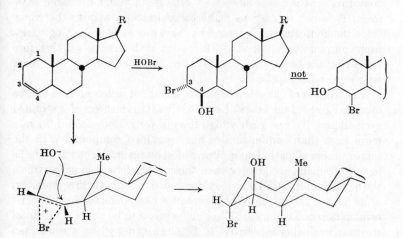

Control of product structure and stereochemistry. Addition of Br⁺ to the less hindered bottom face followed by attack by OH⁻ at C-4 from above puts the substituents diaxial on a chair-shaped ring

The opening of epoxides by nucleophiles is analogous to the opening of cyclic bromonium ions. The structure and stereochemistry of the product depend on the preferential formation of the chair conformation with the substituents axial. Thus in the example below ring A can only become a chair if the bromide ion attacks C-2.

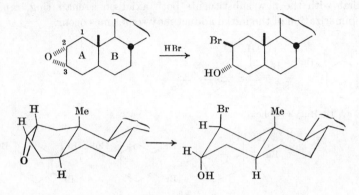

Rings of Other Sizes

Six-membered rings have been discussed in detail since they are more common, for example in natural products, than rings of other sizes, and analysis of their behaviour and properties in terms of conformations is easier since they usually all adopt the same chair form. It is not so easy to make generalizations about other rings since the conformation adopted by, say, a cyclopentane ring varies from compound to compound. Discussion of these rings will therefore be less detailed and we shall treat them together and compare them with one another.

An estimate of the relative strain in rings of different sizes can be obtained from their heats of combustion. Combustion of a strained cycloalkane, i.e. one with a high heat of formation, should liberate more heat than combustion of an unstrained compound with the same number of methylene groups. Since, however, the bond strengths in cyclopropane are greater than those in cyclohexane or paraffins, the heat of combustion of cyclopropane does not give a direct measure of its heat of formation. When corrections are made for such factors, combustion measurements allow an estimate to be made of the total strain in cycloalkanes relative to acyclic alkanes which are assumed to be strain-free. The heat of combustion per methylene group of cyclohexane is the same as that of an alkane, so cyclohexane is strain-free. All other rings have higher heats of combustion per methylene group, and the estimated total strain per mole is given below.

Ring size (n)	3	4	5	6	7	8	9	10	11	12	13	14
Estimated total strain												
(kJ mole^{-1})	190	125	29	0	25	42	54	50	46	17	21	0
(kcal mole^{-1})	45	30	7	0	6	10	13	12	11	4	5	0

It is convenient to group rings according to size and total strain. Thus small rings ($n = 3$ and 4) have considerable strain, normal rings ($n = 5$, 6, and 7) have little or no strain, medium rings ($n = 8$ to 11) are again more strained, and large rings ($n = 12$ or more) are almost strain-free. Let us consider some examples and discuss the factors responsible for the variations in strain.

The cyclopropane ring is necessarily planar. The 60° C–C–C angle implies high angle strain, and this is the major factor affecting the chemical reactions of cyclopropanes. Cyclobutane need not be flat. The actual slightly folded conformation is a compromise between angle strain and torsional strain. Total eclipsing of the CH_2 groups is reduced by puckering at the expense of an increase in angle strain.

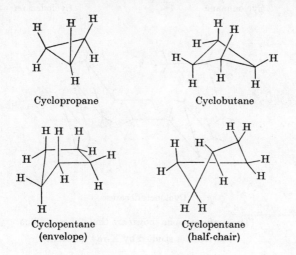

Cyclopropane

Cyclobutane

Cyclopentane (envelope)

Cyclopentane (half-chair)

The angle in a regular pentagon is 108°, so if the ring of cyclopentane were flat there would be almost no angle strain (tetrahedral angle is $109\frac{1}{2}°$) but there would be five fully eclipsed methylene groups causing torsional strain of about 60 kJ mole^{-1} (14 kcal mole^{-1}). To relieve this, the ring is puckered slightly into either an envelope form with one carbon atom out of the plane of the other four, or into a half-chair with two carbon atoms out of the plane of the other three. There are a large number of such flexible forms which are readily interconvertible. This is true also in the case of cycloheptane and cyclooctane and the flexible forms of cyclohexane. In these

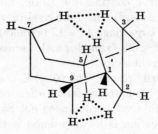

Cyclononane

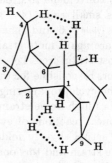

Cyclodecane

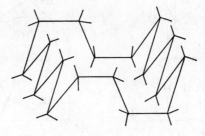

Cyclooctadecane

(The conformations shown are those of crystalline
derivatives studied by X-ray diffraction)

cases the introduction of different substituents may change the most stable conformation. Cyclononane and cyclodecane derivatives have been studied in the solid state by X-ray diffraction. In the crystal the angles in the cyclononanes studied are 117°, and the torsion angles between adjacent methylene groups are about 50°. There is thus some angle strain and some torsional strain and also some steric strain because of interactions between hydrogen atoms in positions 1, 3, and 7 and in positions 2, 5, and 9. Cyclodecanes have angles of 115°, some torsional strain, and steric interactions between hydrogens at positions 1, 4, and 7 and 2, 6, and 9.

Bigger rings seem to adopt conformations as near to those of alkanes as possible with two parallel zig-zags and irregularities only at the ends (see p. 180).

Physical and Chemical Properties of Cycloalkane Derivatives of Different Ring Sizes

(a) *NMR spectra*

We have seen (p. 156) that the coupling constants between vicinal protons depend on the torsion angle between them, and that in rigid cyclohexanes we can often assign a proton to an axial or equatorial position on the basis of experimentally determined coupling constants. Cyclobutanes, cyclopentanes, cycloheptanes, etc. may adopt any of a wide variety of conformations in which torsion angles between vicinal protons can vary considerably.

The variations in torsion angles between substituents on a cyclopentane are shown below as the ring changes from one twist form to another.

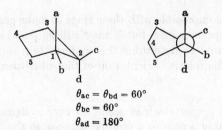

$$\theta_{ac} = \theta_{bd} = 60°$$
$$\theta_{bc} = 60°$$
$$\theta_{ad} = 180°$$

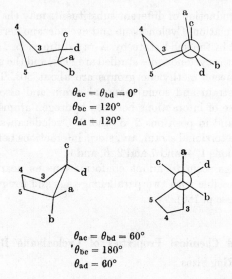

$$\theta_{ac} = \theta_{bd} = 0°$$
$$\theta_{bc} = 120°$$
$$\theta_{ad} = 120°$$

$$\theta_{ac} = \theta_{bd} = 60°$$
$$'\theta_{bc} = 180°$$
$$\theta_{ad} = 60°$$

On the basis of these angles $J_{ac}(cis)$ could be 3—10 Hz, $J_{bc}(trans)$ 0—16 Hz, and $J_{ad}(trans)$ 0—16 Hz. Observed values of both *cis* and *trans* coupling constants in cyclopentanes all fall in the range 0—8 Hz. An almost identical situation occurs in cyclobutanes. The torsion angles depend on the degree and direction of puckering of the ring. Observed values of *cis* and *trans* coupling constants all fall in the range 2—11 Hz.

It is therefore impossible with these rings to make generalizations as we did for cyclohexanes, but it may still be possible to use the coupling constants to determine the conformation in a particular case in which the relative orientation of the substituents is known.

(b) *IR spectra*

One physical property which seems to vary systematically with ring size, probably owing to a gradual increase in the C–C–C angle from cyclopropane to cyclononane, is the infrared carbonyl absorption frequency of the corresponding ketones.

Ring size	3	4	5	6	7—9	acetone
Wave number of carbonyl band (cm^{-1})	1815	1790	1750	1715	1702	1720

(c) *Equilibria and rates of reactions involving sp³–sp² hybridization changes*

The equilibrium constants for the conversion of cycloalkanone cyanohydrins into the corresponding ketones relative to the constant for cyclohexanone are given in the table. The equilibrium lies further to the right, that is more ketone is present at equilibrium, for cyclopentanone and medium and large rings than for cyclohexanone, but further to the left for cyclopropanone.

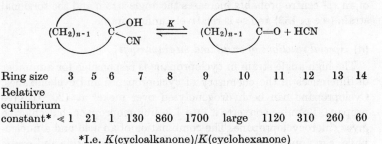

Ring size	3	5	6	7	8	9	10	11	12	13	14
Relative equilibrium constant*	≪ 1	21	1	130	860	1700	large	1120	310	260	60

*I.e. K(cycloalkanone)/K(cyclohexanone)

Similarly, in the S_N1 solvolysis of cycloalkyl toluene-*p*-sulphonates the rate of formation of the carbonium ion is greater for four- and five-membered and medium sized rings than for cyclohexyl, and is very low for cyclopropyl.

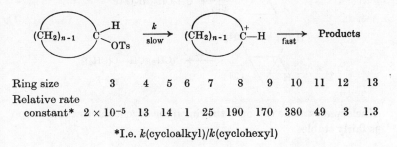

Ring size	3	4	5	6	7	8	9	10	11	12	13
Relative rate constant*	2×10^{-5}	13	14	1	25	190	170	380	49	3	1.3

*I.e. k(cycloalkyl)/k(cyclohexyl)

We must conclude that the change from sp^3 to sp^2 hybridization is very unfavourable in a three-membered ring and is more favourable in 4, 5, and 7—11 membered ones than in cyclohexanes. In other words there is a greater increase in the free energy on going from cyclopropanone cyanohydrin to cyclopropanone than from cyclohexanone cyanohydrin to cyclohexanone, and there is a greater free energy of activation for ionization of the cyclopropyl ester than for the cyclohexyl ester. In the cyclopropane derivatives, introduction of an sp^2 centre increases the already high angle strain and requires an increase in free energy. In the cyclobutane and cyclopentane derivatives, introduction of an sp^2 centre reduces torsional strain, and in the medium sized rings it reduces torsional, steric, and angle strain since the 120° angle is readily accommodated in a ring whose C–C–C angles are constrained to be greater than the tetrahedral angle of $109\frac{1}{2}°$. In the cyclohexane derivatives, introduction of an sp^2 centre probably increases the angle strain and the torsional strain (see p. 153) and so is relatively unfavourable.

(d) *Special reactions of small and medium rings*

The high angle strain in cyclopropane is responsible for a number of differences in the chemistry of cyclopropanes and cyclohexanes. Cyclopropane can be hydrogenolysed over nickel at 120° to give propane. Bromine converts it into 1,3-dibromopropane, but chlorine gives chlorocyclopropane. The combination of an acid and a nucleophile, e.g. aqueous HBr, HI, H_2SO_4, or boron trifluoride and acetic acid opens cyclopropanes. If the ring is unsymmetrically substituted by alkyl groups, cleavage occurs between the least and the most substituted ring atoms.

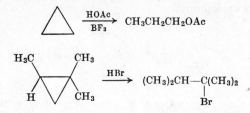

Nucleophiles alone can open the ring if the acyclic intermediate anion is fairly stable.

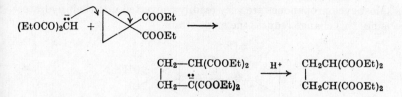

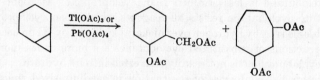

Oxidative cleavages also occur:

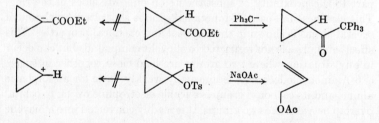

The angle strain would be increased if one carbon atom became sp^2. Therefore cyclopropanecarboxylic esters do not form carbanions.

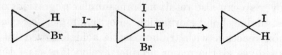

Cyclopropyl cations would also be of very high energy, and unimolecular solvolysis of cyclopropyl tosylate results in ring opening while deamination of cyclopropylamine with aqueous nitrous acid gives allyl alcohol. S_N2 substitution also requires an increase in the C–C–C angle in the transition state, and reaction of cyclopropane derivatives by this pathway is very slow.

Transition state

Most cyclopropanones are very readily converted into their hydrates since this change reduces the angle strain.

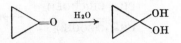

Cyclobutane is less strained than cyclopropane. At room temperature cyclobutane resists all reagents which open cyclopropane but it is cleaved by hydrogen over nickel at 200° and by HI on heating. The reactions of substituted cyclobutanes are more or less normal, and cyclobutanone is not readily converted into its hydrate.

The special conformational situation in medium-sized rings can also lead to unusual chemical behaviour. We have already seen that the conversion of sp^3 to sp^2 hybridization at one carbon atom is favoured in these systems because of the reduction in strain, due in part to the proximity of substituents on opposite sides of the ring. This closeness in space of functional groups separated by chains of several atoms is shown in the physical and chemical properties. This effect, of course, is not restricted to alicyclic compounds but can occur in any situation where two groups are held close together in space.

6-Aminocyclodecanone is unusually weakly basic for an aliphatic amine, and its carbonyl group is a feeble electrophile owing to orbital overlap between the two groups. It is readily converted into a bicyclic enamine.

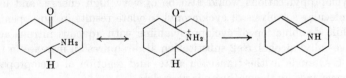

Orbital overlap can also result in transannular migrations of groups or transannular bond formation. For example, *cis*-cyclooctene epoxide on reaction with trifluoroacetic acid followed by alkaline hydrolysis gives as the major product cyclooctane-*cis*-1,4-diol as a result of transannular hydride shifts (both 1,5 and 1,3). The conformation adopted in the reaction is not known but the 1,5 hydride shift can be drawn as shown.

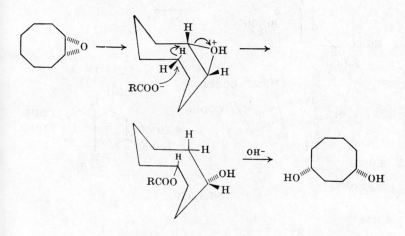

Participation of transannular double bonds in solvolysis (Scheme A) or addition reactions (Scheme B) results in formation of a new bond across the ring. Note that *trans*-addition occurs at both double bonds in the *cis,cis*-cyclodecadiene (Scheme B) and all the new bonds are parallel. The final stereochemistry thus depends on the mechanism of the reaction and on the stereochemistry of the starting materials. In the case of the acid-catalysed cyclization (Scheme C) the final stereochemistry is determined also by the conformation adopted by the starting diene.

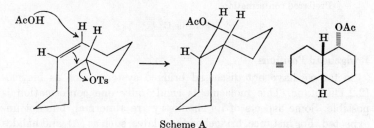

Scheme A

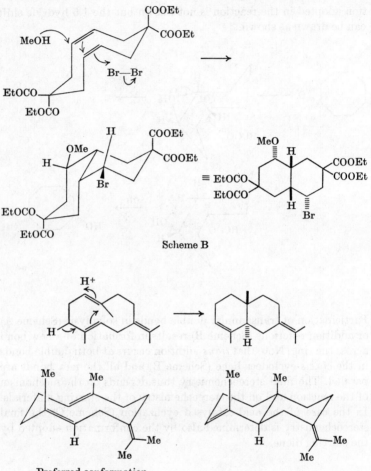

Scheme B

Scheme C

Preferred conformation

Bridgehead Positions

So far we have not discussed bridged systems such as bicyclo-[2,2,1]heptane. This molecule is rigid; only one conformation is possible. Some aspects of its chemistry are abnormal, but not unexpected. For instance, bridgehead tosylates, such as (A), and halides are very stable to nucleophilic attack, since the bridgehead carbon atom cannot become planar as required for S_N1 reactions or invert as required for S_N2 reactions, nor do they undergo elimination reactions. Indeed, double bonds at bridgeheads are impossible unless the bridges are fairly long (Bredt's rule). Thus bicyclo[2,2,1]hept-1-ene (B) has never been made although bicyclo[3,3,1]non-1-ene (C) is known.

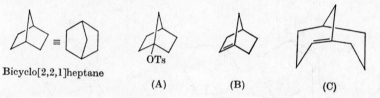

Bicyclo[2,2,1]heptane

(A)   (B)   (C)

For the same reason (D) does not show the properties of a simple enolizable β-diketone, and (E) does not decompose on heating as do other β-keto-acids (Part 2, p. 197).

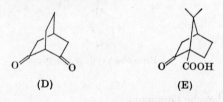

(D)   (E)

Radicals and carbanions can, however, be formed at the bridgehead, and this is in accord with our knowledge that, even in acyclic systems, carbanions and radicals are not, or are not necessarily, planar.

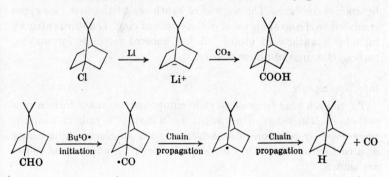

The formation of a cation, an anion, and a radical are, however, all possible at the 1-position in adamantane. The radical is, in fact, more stable than the t-butyl radical. The cation is readily formed in strongly acidic media and is an intermediate in the unimolecular solvolysis of 1-bromoadamantane which is only 1000 times slower than that of t-butyl bromide while 1-bromobicyclo[2,2,1]heptane solvolyses about 10^{14} times slower than t-butyl bromide.

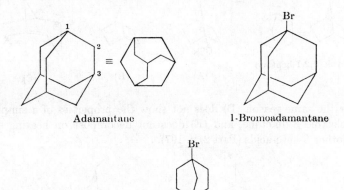

Adamantane   1-Bromoadamantane

1-Bromobicyclo[2,2,1]heptane

Synthesis of Alicyclic Compounds

The remainder of this chapter discusses the synthesis of alicyclic compounds under three headings: formation of rings by closing a chain, formation of the required ring from another ring, and formation by special methods. The success of syntheses of the first two types is related to the strain present in the required ring. The third category includes a variety of useful but less general methods for making carbocyclic rings of different sizes.

(a) *Ring closure*

To make a ring from an acyclic compound we must form a new carbon–carbon bond. This might be done by a radical coupling process or by reaction of a carbanion with an electrophilic centre or a carbonium ion with a nucleophilic centre. Some general routes are shown.

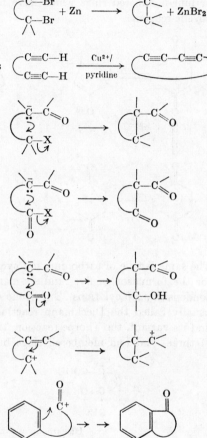

Wurtz type reductive coupling

Oxidative coupling of acetylenes

Alkylation

Acylation

Aldol condensation

Carbonium ion cyclization

Friedel–Crafts aromatic
substitution.

Whether any of these reactions gives the desired cyclized product in good yield depends on several factors. If the reaction is reversible the amount of product formed depends on the value of the equilibrium constant, that is on the relative stabilities of product and starting material. If several products are possible, the one with the lowest free energy under the conditions will be formed predominantly.

Aldol condensations are completely reversible. They are useful

for the synthesis of five- and six-membered rings but fail for three-
or four-membered rings which have higher free energies than the
acyclic reagent.

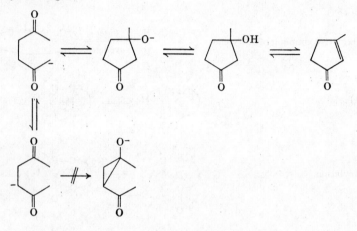

The same is true of carbonium ion cyclizations, which succeed only
for the formation of five- and six-membered rings, and of Claisen
condensations (acylations). The cyclic version of the latter, which is
usually called the Dieckmann reaction (Part 2, pp. 99 and 199)
and its variant, the Thorpe reaction, therefore fail for the synthesis
of three- and four-membered rings but are very useful for larger

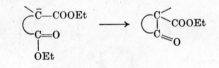

Dieckmann cyclization

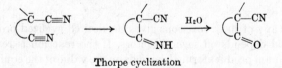

Thorpe cyclization

rings. However, the yields are poor for rings larger than C_8 since the
ends of the chain are so remote from one another and polymerization
competes favourably with cyclization.

The chances of an anion cyclizing rather than condensing with a second ester molecule are increased if there are no other ester molecules around, i.e. if the reaction is done in dilute solution in an inert solvent. By carrying out the Thorpe reaction in dilute solution, the yields of the strain-free big rings (>C_{14}) can be raised to about 50%. The yields of the strained medium-sized rings however remain very low and they can only be made satisfactorily by irreversible reactions.

If the reaction we try is irreversible, it should succeed if it is the fastest reaction possible under the conditions. The rate of the reaction depends on the free energy of activation, $\Delta G^{\ddagger} = \Delta H^{\ddagger} - T\Delta S^{\ddagger}$. The enthalpy of activation, $\Delta H^{\ddagger}$, will be affected by the amount of strain in the transition state, and the entropy of activation, $\Delta S^{\ddagger}$, will be affected by the probability of the chain ends coming together, i.e. by the change in mobility in the system. The situation can be summarized as in the table.

Ring size	3	6	10
$\Delta H^{\ddagger}$	+ve large	+ve small	+ve large
$-\Delta S^{\ddagger}$	+ve small	+ve medium	+ve large

This suggests that irreversible cyclizations should be most successful for the preparation of three- and six-membered rings. The yields obtained in practice are more or less in accord with this prediction.

Thus, in the reaction of zinc with dihalides, $Br(CH_2)_nBr$, the yields of cycloalkanes, C_nH_{2n}, are 80% for $n = 3$, 7% for $n = 4$, 45% for $n = 6$, and 1% for $n = 10$. In the irreversible pyrolysis of heavy-metal salts of organic dicarboxylic acids $HOCO(CH_2)_{n-1}COOH$, to give cycloalkanones, $(CH_2)_{n-1}$—C=O, and the metal carbonate, the yields vary somewhat with the metal used but are of the order shown:

n	5	6	7	8	9	10—12	16
%	50—70	50—70	40—50	5—20	2	0	10

The best ring closure method of making medium and large rings is the reduction of diesters with sodium to give ketols or acyloins, known as the acyloin reduction. It is usually performed by adding the diester to a suspension of droplets of molten sodium in hot toluene. The polar ends of the chains are brought together on the surface of the droplets, so that the entropy barrier is reduced. Yields of medium rings are good (60—70%) and of large rings excellent (90%).

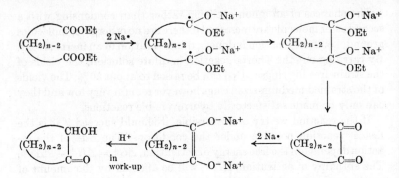

By performing the reaction in homogeneous solution in liquid ammonia yields of 50—60% can be obtained for normal-sized rings.

(b) *Changes of ring size*

It is often convenient to make a ring from another available one which is a little smaller or a little larger. Some examples of this are shown. Most of them involve molecular rearrangements of types we have discussed before.

Favorskii rearrangement. Several mechanisms are known to operate in different cases. Two possible routes are shown.

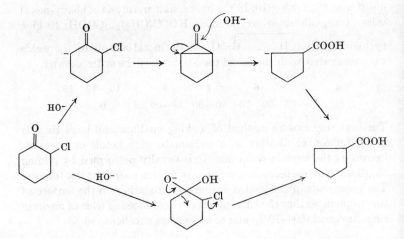

Carbene–ketene rearrangement (cf. Part 2, p. 176).

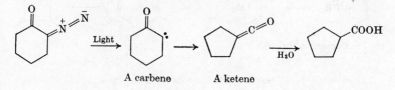

A carbene   A ketene

Diazomethane addition (cf. Part 2, p. 177).

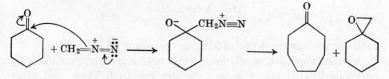

Carbonium ion rearrangements (cf. Part 2, pp. 113 and 116)

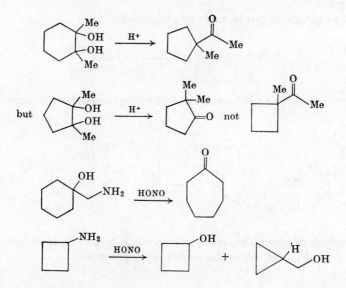

(c) *Special synthetic methods*

There are, of course, methods of making cyclic compounds other than ring closure by formation of one new bond. The formation of two new bonds occurs in cycloaddition reactions such as those shown.

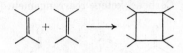

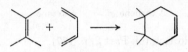

Carbene addition

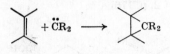

Diels–Alder addition

Rules governing these cycloadditions are discussed in Chapter 7.

Another approach would be to start with a readily available ring compound and convert it into the one required. These and other special methods of constructing rings will be discussed in order of ring size.

Cyclopropanes can be made by cyclization.

$$\text{PhCHBrCH}_2\text{CH}_2\text{Br} \xrightarrow{\text{Zn/Cu}} \text{Ph-}\triangle$$

They can also be made from enones by heating the derived pyrazolines.

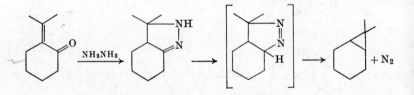

Pyrolysis of pyrazolines made by a cycloaddition of diazoalkanes to electron-deficient olefins also leads to cyclopropanes.

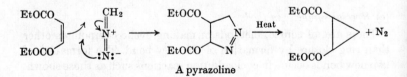

A pyrazoline

Reaction of olefins with methylene iodide and copper-coated zinc, the Simmons–Smith reaction, gives cyclopropanes. The mechanism is not definitely known.

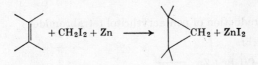

Similar additions of a carbon atom to a double bond can be effected by using carbenes (see Part 2, p. 285). The carbenes are obtained from diazoalkanes or by base-induced α-eliminations.

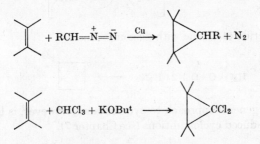

Electron-deficient olefins react with ylids (cf. Chapter 5, p. 233) made from dialkyl sulphoxides to give cyclopropanes. The ylids from phosphines or phosphonates react with epoxides to replace the oxygen by carbon. The reactions can be represented as shown.

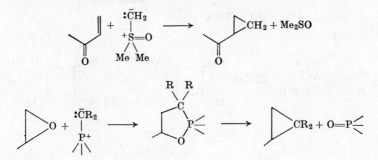

Cyclobutanes can be made conveniently by the reaction of diethyl malonate with 1,3-dibromopropane in base (Part 2, p. 195):

$$Br(CH_2)_3Br + CH_2(COOEt)_2 + EtO^- \longrightarrow$$ (cyclobutane with COOEt, COOEt)

or by zinc reduction of pentaerythritol tetrabromide:

$$BrCH_2{-}\underset{\underset{CH_2Br}{|}}{\overset{\overset{CH_2Br}{|}}{C}}{-}CH_2Br + Zn \longrightarrow$$

or by reaction of ketene with diazomethane:

$$H_2C{=}C{=}O + 2\ CH_2N_2 \longrightarrow \quad + N_2$$

The most general approach to cyclobutanes, however, is by thermal or photoinduced cycloadditions (see Chapter 7).

$$2\ H_2C{=}C{=}CH_2 \xrightarrow{500°}$$

$$H_2C{=}C{=}CH_2 + H_2C{=}CHCN \xrightarrow{200°}$$

$$2\ \underset{H}{\overset{Ph}{>}}C{=}C\underset{COOH}{\overset{H}{<}} \xrightarrow{Light}$$

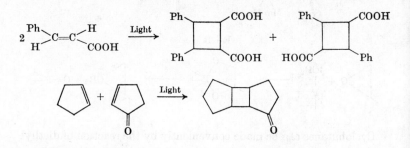

Five- and six-membered rings are readily available by aldol and Dieckmann cyclizations. A similar reversible carbanion condensation called the Robinson–Michael reaction provides a useful route to cyclohexenones.

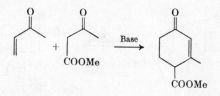

Diels–Alder cycloaddition (Part 1, p. 124; Part 2, p. 348) produces cyclohexenes:

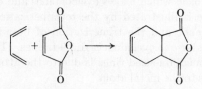

and catalytic hydrogenation or chemical reduction of benzene derivatives is a convenient route to various cyclohexane derivatives.

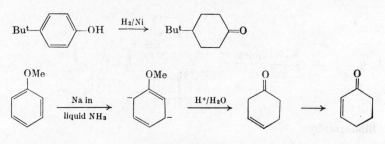

Birch reduction

Larger rings can be made by ring closure methods described above but there are some special methods such as the opening of bicyclic molecules. This is particularly useful for making nine- and ten-membered rings since the required [4,4,0] and [4,3,0] systems are easily obtained.

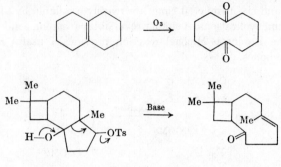

Fragmentation

A different approach which makes cyclooctanes and cyclododecanes readily available is illustrated by the tetramerization of acetylene to cyclooctatetraene and the trimerization of butadiene to *trans,-trans,cis-* and all-*trans*-cyclododeca-1,5,9-trienes. The success of these routes to medium-sized rings is due to the attachment of the olefins as ligands to the metal atom.

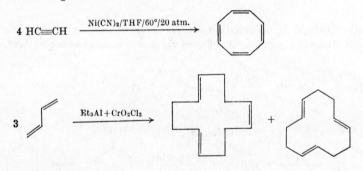

Bibliography

Recent reviews of conformational analysis

E. L. Eliel, N. L. Allinger, S. J. Angyal, and G. A. Morrison, *Conformational Analysis*, Interscience, New York, 1965.

M. Hanack, *Conformation Theory*, Academic Press, New York, 1965.

E. L. Eliel, *Stereochemistry of Carbon Compounds*, McGraw-Hill, New York, 1962.

J. McKenna, *Conformational Analysis*, Lectures, Monographs and Reports, Royal Institute of Chemistry, No. 1, 1966.

J. A. Hirsch, "Conformational Energies," in *Topics in Stereochemistry* (Ed. N. L. Allinger and E. L. Eliel), Vol. 1, 1967, p. 199.

General background material

K. Mislow, *Introduction to Stereochemistry*, Benjamin, New York, 1966.
R. O. C. Norman, *Principles of Organic Synthesis*, Methuen, London, 1968.
G. H. Whitham, *Alicyclic Chemistry*, Oldbourne, London, 1963.
M. S. Newman (Ed.), *Steric Effects in Organic Chemistry*, Wiley, New York, 1956.
S. Coffey (Ed.), *Rodd's Chemistry of Carbon Compounds*, 2nd Edn, Vols. IIA and IIB, Elsevier, Amsterdam, 1968.

Physical methods

N. S. Bhacca and D. H. Williams, *Applications of NMR Spectroscopy in Organic Chemistry*, Holden-Day, San Francisco, 1964.
P. Crabbé, *Optical Rotatory Dispersion and Circular Dichroism in Organic Chemistry*, Holden-Day, San Francisco, 1965.
L. Velluz, M. Legrand, and M. Grosjean, *Optical Circular Dichroism*, Verlag Chemie/Academic Press, 1965.
G. Snatzke, *Angew. Chem.* (International Edn. in English), **7**, 14 (1968).
P. Crabbé, "Recent Applications of ORD and CD in Organic Chemistry," in *Topics in Stereochemistry* (Ed. N. L. Allinger and E. L. Eliel), Interscience, New York, 1967, p. 93.

Specialized articles

D. L. Robinson and D. W. Theobald, "Conformational Abnormalities in Cyclohexane Chemistry," in *Quarterly Reviews*, **21**, 314 (1967).
F. Johnson, "Allylic Strain in Six-membered Rings" (cyclohexenes and alkylidenecyclohexanes), in *Chemical Reviews*, **68**, 375 (1968).
R. C. Fort and P. von R. Schleyer, "Bridgehead Reactivity", in *Advances in Alicyclic Chemistry*, Vol. 1, 1966, p. 283.
R. Breslow, "Rearrangements in Small Ring Compounds," in *Molecular Rearrangements* (Ed. P. de Mayo), Interscience, New York, 1964.
V. Prelog and J. G. Traynham, "Transannular Hydride Shifts," in *Molecular Rearrangements* (Ed. P. de Mayo), Interscience, New York, 1964.
A. C. Cope *et al.*, "Transannular Reactions in Medium Sized Rings," in *Quarterly Reviews*, **20**, 119 (1966).
C. D. Gutsche and D. Redmore, "Carbocyclic Ring Expansion Reactions," Supplement No. 1 to *Advances in Alicyclic Chemistry* (Ed. H. Hart and G. Karabatsos), Academic Press, 1968.
R. A. Pethrick and E. Wyn-Jones, "Determination of the Energies Associated with Internal Rotation," in *Quarterly Reviews*, **23**, 301 (1969).

Problems

1. If (A) represents one possible conformation of (+)-*cis*-1-bromo-2-methylcyclohexane, what do the other drawings represent?

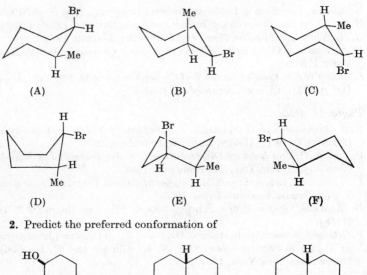

(A)   (B)   (C)

(D)   (E)   (F)

2. Predict the preferred conformation of

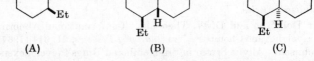

(A)   (B)   (C)

3. Draw conformational representations of each of the following and discuss the possibility of resolving them at room temperature:

cis-1,2-, *trans*-1,2-, and *cis*-1,3-dichlorocyclohexane.

4. In non-polar solvents the preferred conformations of 2-bromocyclohexanone, *trans*-1,2-dibromocyclohexane, *cis*-1,3-dihydroxycyclohexane, and *trans*-1,3-di-t-butylcyclohexane are as shown. Suggest reasons for this.

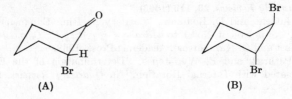

(A)   (B)

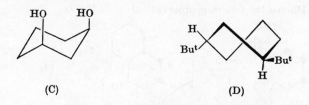

(C) (D)

5. How do you account for the course of the following reactions?:

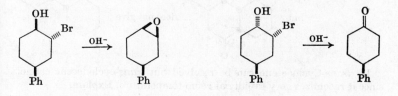

6. Suggest why *cis*-4-t-butylcyclohexanecarboxylic acid is a weaker acid than its *trans*-isomer.

7. Account for the observed product of the reaction:

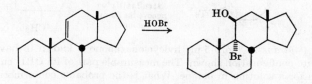

8. On the basis of the figures in the table on p. 149, and assuming that there is no entropy change, calculate the percentage of each in a mixture of *cis*- and *trans*-4-t-butylcyclohexanol equilibrated over nickel at 127°.

9. (a) How many stereoisomeric forms of each of the following are possible?

(b) Which isomer do you expect to be the most stable in each case?

(c) Which of the compounds might be obtained in an optically active form at room temperature?

(i) Bicyclo[3,2,1]octane; (ii) bicyclo[3,3,0]octane;

(iii) tricyclo[8,4,0,0³,⁸]tetradecane; (iv) bicyclo[2,2,2]octan-2-ol.

10. Discuss the following observations:

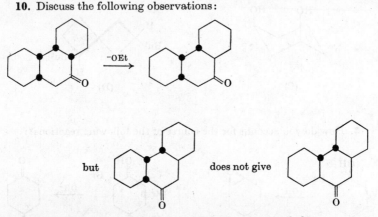

11. *trans*-Cyclooctene can be resolved but *trans*-cyclodecene cannot since it racemizes very rapidly at room temperature. Explain.

12. Compared with those of simple olefins, the heats of hydrogenation (−*ΔH*) of *trans*-cyclooctene and *cis*-cyclodecene are very high and very low respectively. Why?

13. How would you account for the following observation?:

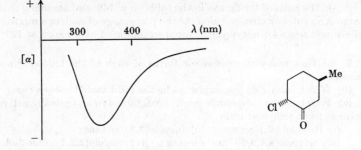

14. (−)-*trans*-2-Chloro-5-methylcyclohexanone is known to have the absolute configuration shown. The measurable part of its ORD curve is shown for a solution in octane. What is the probable conformation in this solvent?

CHAPTER 4

Saturated Heterocycles

The chemistry of aromatic heterocycles, where conformation and strain are relatively unimportant, has been discussed in Chapter 2. The importance of these factors in the flexible saturated alicyclic compounds has been discussed in Chapter 3. We can now extend this discussion to the chemistry of saturated heterocyclic compounds. The names and structures of the more common of these are shown.

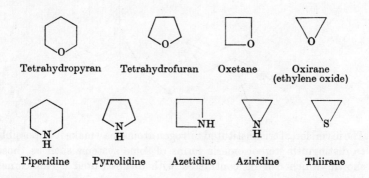

| Tetrahydropyran | Tetrahydrofuran | Oxetane | Oxirane (ethylene oxide) |

| Piperidine | Pyrrolidine | Azetidine | Aziridine | Thiirane |

Conformation

We might expect such rings to be subject to the same conformation-determining factors as carbocyclic rings with, also, possible effects due to the heteroatom. For example, C–O bonds are shorter than C–C bonds; therefore, in tetrahydropyran C-1 and C-5 are closer together than they are in cyclohexane.

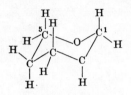

Tetrahydropyran

Similarly, the optimum C–O–C bond angle may be different from the C–C–C angle. Moreover, the heteroatom carries one or two lone pairs of electrons in place of the bonding pairs on carbon atoms. The lone pairs may have different steric and electrostatic requirements from bonding pairs, so barriers to rotation and torsional strains may be different from those in the carbocyclic analogues.

The dipoles of the C–O ring bonds may be important. For example, equatorial 2-chlorotetrahydropyrans are less stable than their axial epimers owing to an unfavourable interaction of the oxygen and chlorine dipoles which are more closely aligned in the former.

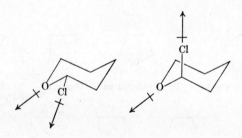

The inversion of trisubstituted nitrogen atoms may make it impossible to distinguish stereoisomeric forms of some systems such as those shown, which can be contrasted with disubstituted cyclohexanes and decalins (see Chapter 3).

Quinolizidine

Two *conformations* of these heterocycles are possible, and the rate of interconversion will depend on the structure of the molecule and the temperature. In the carbocyclic analogues the two forms are distinct and separable configurational isomers.

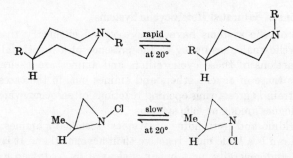

It is indeed found that tetrahydropyran and piperidine rings exist in chair forms, and it has been estimated that these are about 20 kJ mole^{-1} (5 kcal mole^{-1}) more stable than the boat or twist forms. Tetrahydropyran rings are very common in sugars (see Part 1, p. 157). On the basis of our experience with cyclohexanes we should predict that methyl β-D-glucopyranoside should exist in the C_1^4 or 'normal' or C1 form, rather than the C_4^1 or 'alternative' or 1C form, and it does.

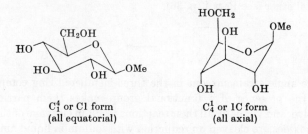

C_1^4 or C1 form
(all equatorial)

C_4^1 or 1C form
(all axial)

Piperidine rings are common in the complex naturally occurring amines, the alkaloids, and are generally found to adopt chair conformations. Tetrahydrofuran and pyrrolidine have puckered rings with many readily interconvertible conformations, as in the case of cyclopentane. In oxetane and azetidine the rings are believed to be flat and there is some strain due to angle deformation. In the planar oxiranes, thiiranes, and aziridines this angle strain is the controlling factor in their chemical reactivity and the total strain in these three systems has been estimated to be about 54, 38, and 59 kJ mole^{-1} (13, 9, and 14 kcal mole^{-1}) respectively, compared with 110 kJ mole^{-1} (27 kcal mole^{-1}) or more for cyclopropane. The reasons for this apparent difference in strain are not clear.

Reactions of Saturated Heterocyclic Systems

Heterocyclic systems have a more varied chemistry than their carbocyclic analogues owing to the presence of a functional group. The reactions of these cyclic ethers and amines are basically the same as those of acyclic ethers and amines but, in the cases where angle strain is great, ring opening reactions often occur which have no analogues among acyclic systems.

Piperidine and pyrrolidine are typical secondary amines. Tetrahydrofuran is a little more reactive than acyclic ethers. It is a good solvent and generally inert but it is cleaved by hydrogen iodide at room temperature to 1,4-diiodobutane, by acetyl chloride at 65° to 4-chlorobutyl acetate, and by hydrogen over platinum at 250° or by lithium aluminium hydride and aluminium chloride together to butan-1-ol.

Oxetane can be cleaved by lithium aluminium hydride alone and by nucleophiles in the presence of acids. That is, as we should expect, the oxonium ion is more readily attacked by nucleophiles than the neutral ether (see Part 1, p. 36).

The angle strain present in the three-membered ring compounds plus the presence of a functional group makes them particularly reactive, and virtually all their reactions result in cleavage of the ring.

Epoxides are cleaved on reduction with sodium in liquid ammonia or hydride donors to give, in most cases, the more substituted alcohols:

but catalytic hydrogenation gives both possible alcohols. Reduction with hydrogen iodide gives olefins and alkanes, while phosphines and phosphites give the related olefin (cf. Chapter 5).

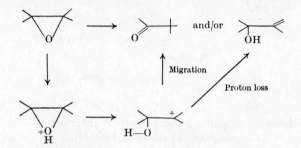

On treatment with acids such as H^+, $MgCl_2$, or BF_3 in the absence of nucleophiles, epoxides tend to rearrange either to carbonyl compounds or to allylic alcohols; the mechanism involves cleavage of the oxonium ion followed either by migration of a substituent or by proton loss.

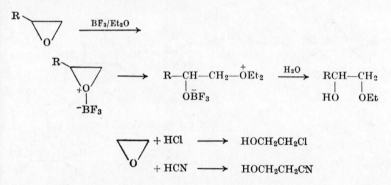

In the presence of nucleophiles the oxonium ion is usually attacked by them as in the following examples:

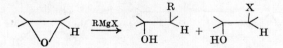

Treatment with Grignard reagents tends to give mixtures of alcohols and halohydrins:

The magnesium halide present may also catalyse an initial rearrangement to a ketone which then reacts with RMgX. Ring opening by nucleophiles may take different paths depending on the presence or absence of an acid:

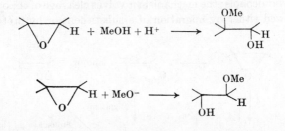

In all these nucleophilic attacks there is inversion at the centre attacked. This stereochemical aspect is illustrated by the sequence of reactions which results when the nucleophile is thiocyanate ion.

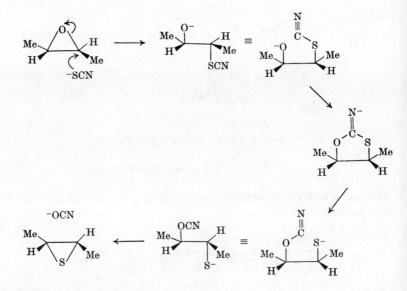

The initial *trans*-oxirane gives a *trans*-thiirane of opposite absolute configuration. Since all the steps are reversible, the greater stability of the thiirane (see above) decides the outcome. The initial ring opening with inversion is followed by a conformational change and transfer of the cyano group from S to O. Re-formation of a ring, again with inversion, gives the *trans*-thiirane.

Carbanion nucleophiles react, like methoxide, at the less substituted end. The initial adduct often cyclizes to a lactone.

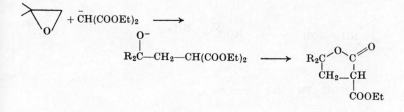

Note, however, that cyclopentene oxide does not yield a lactone.

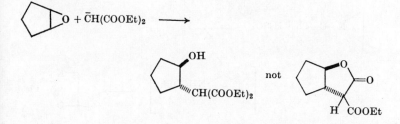

Ring opening again occurs with inversion. Formation of a lactone would mean the *trans*-fusion of two five-membered rings which would cause a considerable increase in strain.

Aziridines are secondary amines but, particularly when the nitrogen is positively charged, they are very prone to ring cleavage. Their *N*-alkyl and *N*-acyl derivatives can be prepared normally. The ring is fairly stable to reduction in basic media.

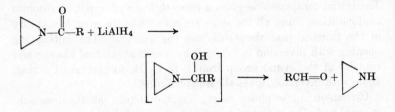

Simple ring opening by nucleophiles occurs as with epoxides, resulting in relief of ring strain.

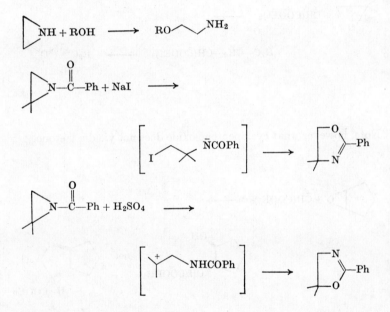

Note again that the acid-catalysed and non-acid-catalysed reactions give different products.

Preparations of Heterocycles by Chain Closing

The ease of ring closure by a reversible process is very dependent on ring size. Thus acid-catalysed lactonization of hydroxy- or unsaturated acids to give δ-lactones is easy, and to γ-lactones often spontaneous, while β- and α-lactones cannot be prepared in this way.

Similarly, acid-catalysed cyclizations of hydroxy-olefins occur readily to give five- or six-membered rings but not four- or three-membered ones.

Irreversible reactions may succeed for all ring sizes provided that there is no competing alternative reaction. Some ring closure reactions involving nucleophilic displacements are shown. Some examples of these and related routes are given in Chapter 18 of Part 2.

Type A (heteroatom as nucleophile)

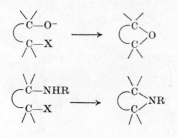

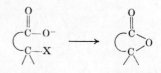

Type B (carbon as nucleophile)

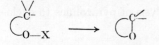

Other Preparative Methods

As with the carbocyclic rings, special methods of ring formation other than chain cyclization are often more convenient in certain cases but may not be generally applicable. Piperidine and tetrahydrofuran and their derivatives are often prepared by hydrogenation of the more readily available aromatic parents. The four-membered rings can often be prepared by cycloaddition reactions induced by heat or light.

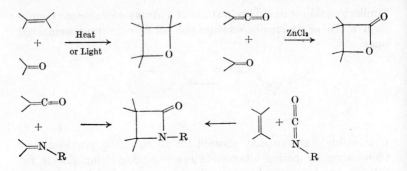

The thermodynamically unstable small rings can often be prepared by pyrolytic elimination of a very stable small molecule from some stable precursor.

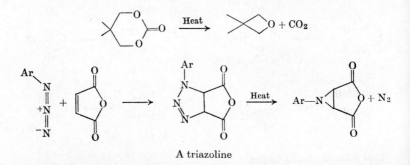

A triazoline

The latter reaction resembles the pyrolysis of pyrazolines to cyclopropanes (see page 196).

Preparation of Epoxides (Oxiranes)

Epoxides can be made by ring closure by the routes outlined above (p. 213) and also by some special methods. The dehydrohalogenation of halohydrins (type A, p. 213) has been exemplified in Chapter 3, p. 173.

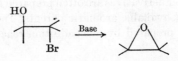

Mechanistically similar are the conversions of carbonyl compounds into epoxides with ylids. Thus ketones and aldehydes react with diazomethane to give epoxides and higher ketones (see Part 2, p. 177).

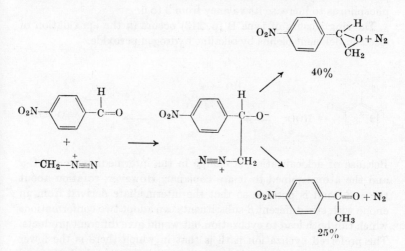

40%

25%

Sulphur ylids (see Chapter 5) react with carbonyl groups in a similar way (p. 236).

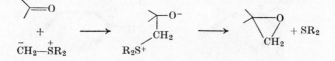

With simple ketones, the ylids made from dialkyl sulphides and those from dialkyl sulphoxides behave analogously, but with enones different products are obtained in the two cases.

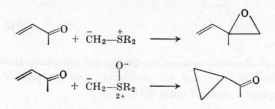

It is interesting to compare this reaction with the reaction of carbonyl compounds with phosphorus ylids, $R_2\bar{C}–\overset{+}{P}R'_3$ (the Wittig reaction, p. 234). The difference can be attributed to the greater tendency of phosphorus to increase its valency from 3 to 5.

The ring closure of type B (p. 213) occurs in the epoxidation of electron-deficient olefins by alkaline hydrogen peroxide.

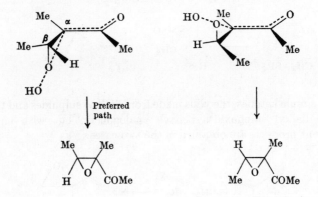

Because of delocalization of charge in the intermediate anion, C-α and the atoms joined to it are coplanar. However, rotation about the α–β bond is possible so that the intermediate derived from an enone with two different β-substituents can adopt two conformations which can both lead to cyclization but would give different products. The preferred cyclization path is that in which there is the lower steric strain in the transition state, namely, that in which the acyl group is eclipsed with the β-H rather than the β-alkyl group.

The product observed is therefore the epoxide in which the acyl and the β-methyl are *trans*, and this product is found irrespective of which of the isomeric enones we start with.

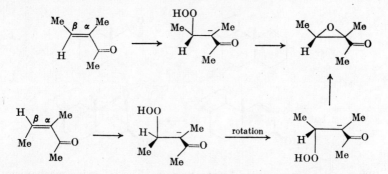

It is possible that in other cases cyclization could be faster than rotation, in which case the reaction product would retain the stereochemistry of the starting material.

The direct addition of an oxygen atom to an olefin is another important route to epoxides. Ethylene is oxidized by oxygen on the surface of a silver catalyst to ethylene oxide in about 50% yield. Some inorganic oxidizing agents such as chromium trioxide and pertungstic acid also oxidize olefins to oxiranes.

A more general method is the use of peracids such as peracetic, trifluoroperacetic, and substituted perbenzoic acids (Part 1, p. 58).

The reaction is stereospecific in that *cis*-olefins give *cis*-disubstituted oxiranes.

The oxygen usually adds to the less hindered face of the olefin. However, if there is a hydroxy group near the double bond it may control the reaction so that the OH and epoxide are *cis*. This is presumably due to hydrogen-bonding with the reagent. These two aspects are illustrated by the reactions of the steroids shown (p. 218).

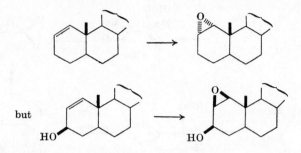

The reaction is catalysed by strong acids. The following mechanism is in accord with the known facts.

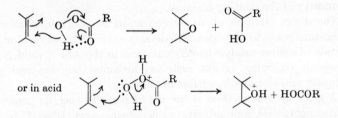

An alternative to peracids is the combination of benzonitrile and alkaline hydrogen peroxide, which gives epoxides and benzamide presumably by an analogous mechanism.

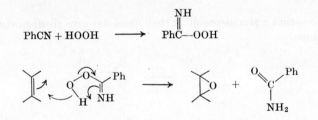

Since these reactions are initiated by nucleophilic attack by the olefin on the peracid, they give poor yields from electron-deficient olefins. Such olefins are, however, readily epoxidized by the hydrogen peroxide method discussed above.

Preparation of Aziridines

In addition to the ring closure methods involving internal displacement, methods have been developed for the formation of aziridines by addition of a nitrogen atom to a carbon–carbon double bond. The reaction is similar to the carbene–olefin addition approach to cyclopropanes. Photolysis of ethyl azidoformate in the presence of cyclohexene or treatment of arenesulphonyloxyurethanes with triethylamine in the presence of cyclohexene gives the aziridines shown along with a little of the urethane. It is believed that the intermediate in both cases is the nitrene, EtOCON (Part 2, p. 292), formed as shown.

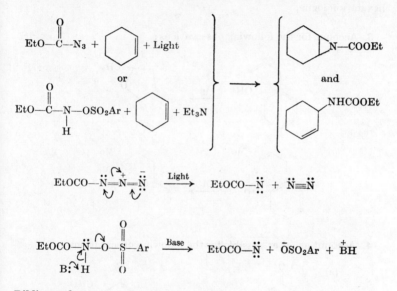

Bibliography

R. M. Acheson, *An Introduction to the Chemistry of Heterocyclic Compounds*, 2nd Edn., Interscience, 1967.

A. Weissberger (Ed.), *Heterocyclic Compounds with Three- and Four-membered Rings*, Interscience, New York, 1964.

M. Hanack, *Conformation Theory*, Academic Press, New York, 1965, Chapter 7.

E. L. Eliel, N. L. Allinger, S. J. Angyal, and G. A. Morrison, *Conformational Analysis*, Interscience, New York, 1965, Chapter 6.

J. McKenna, Royal Institute of Chemistry Lecture Series, 1966, No. 1, p. 70.

Problems

1. The preferred conformation of 5-hydroxy-1,3-dioxan in non-polar solvents is a chair with the hydroxy group axial. Discuss.

　　　　　　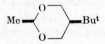

5-Hydroxy-1,3-dioxan　　　　　*cis*-2-Methyl-5-t-butyl-1,3-dioxan

2. The preferred conformation of *cis*-2-methyl-5-t-butyl-1,3-dioxan is a chair with the t-butyl group axial. Discuss and contrast with the cyclohexane analogue.

3. Account for the following observations:

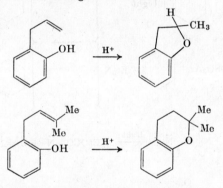

4. Account as far as possible for the following observations:

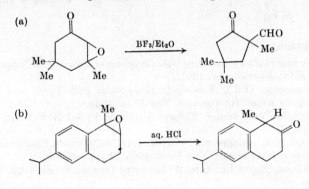

(c)

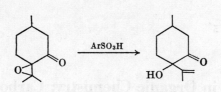

d-Orbitals in Organic Chemistry: Carbon Compounds containing Phosphorus and Sulphur

d-Orbitals

In Chapter 2 of Part 2 we saw that to obtain solutions of the Schrödinger equation for a hydrogen atom we required three quantum numbers: the principal quantum number n (n can be any whole number equal to or greater than 1), the angular quantum number l ($0 \leqslant l \leqslant n - 1$), and the magnetic quantum number m ($-l \leqslant m \leqslant l$). We saw that selection of these quantum numbers led to solutions of the wave equation with specific energy, and each solution defined a probability distribution of the electron. The solutions were called orbitals. The orbital of lowest energy ($n = 1$, $l = m = 0$) we called the 1s orbital. When $n = 2$ there were four solutions of equal energy, the 2s orbital ($n = 2, l = m = 0$) which like the 1s orbital was spherically symmetric, and three 2p orbitals ($n = 2$, $l = 1$, $m = 0$; $n = 2$, $l = 1$, $m = 1$; $n = 2$, $l = 1$, $m = -1$) all of which had a node at the nucleus. We did not consider what happens when $n = 3$. If we do, we find that there are nine solutions of equal energy, i.e. nine degenerate orbitals. The first solution ($n = 3, l = 0, m = 0$) is the 3s orbital and is spherically symmetric. The next three solutions ($n = 3$, $l = 1$) are the 3p orbitals which have a spatial distribution similar to the spatial distribution of the 2p orbitals. We now have five more solutions ($n = 3$, $l = 2$) which are known as 3d orbitals each of which has two nodes and which are depicted on page 223.

These diagrams are polar diagrams (see Part 2, Chapter 2), and should not be regarded as accurately picturing shape. The s and p orbital diagrams are the same as we had before. The d_{z^2} orbital is symmetric around the z axis. The d_{xy}, d_{yz}, and d_{zx} orbitals are exactly

Table 5.1. Orbitals of hydrogen-like atoms

Quantum numbers			General symbol	Explicit symbol
n	l	m		
1	0	0	$1s$	$1s$
2	0	0	$2s$	$2s$
2	1	0		$2p_x$
2	1	1	$2p$	$2p_y$
2	1	−1		$2p_z$
3	0	0	$3s$	$3s$
3	1	0		$3p_x$
3	1	1	$3p$	$3p_y$
3	1	−1		$3p_z$
3	2	0		$3d_{z^2}$
3	2	1		$3d_{xy}$
3	2	−1	$3d$	$3d_{yz}$
3	2	2		$3d_{x^2-y^2}$
3	2	−2		$3d_{xy}$

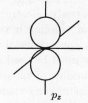
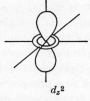
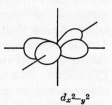

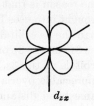

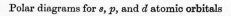

Polar diagrams for s, p, and d atomic orbitals

alike except that they lie in the xy, yz, and zx planes respectively. The $d_{x^2-y^2}$ orbital again has the same shape but it is rotated by 45° around the z axis so that its lobes are directed along the axes.

So far we have been talking about solutions of the Schrödinger equations for the hydrogen atom, and in this case all the orbitals with principal quantum number 2 have the same energy and similarly all the orbitals with principal quantum number 3 have the same energy; that is, they are degenerate. We saw in Part 2, p. 17, that in a system containing more than one electron, interelectronic repulsion occurs and this results in the energy levels predicted for the hydrogen atom becoming further split. Whereas for the hydrogen atom the $2s$ and the three $2p$ orbitals are degenerate, in atoms with more than one electron these separate into one $2s$ orbital which has a slightly lower energy than three degenerate $2p$ orbitals. In the same way the nine orbitals with principal quantum number 3 are split into three levels. The new energies are such that np orbitals have higher energy than ns but lower than $(n + 1)s$ orbitals; nd orbitals have about the same energy as $(n + 1)s$ but are of lower energy than $(n + 1)p$ orbitals.

Atoms with Valence Shells of More than Eight Electrons

In the first row of the Periodic Table elements in Groups IV, V, VI, and VII form compounds such that their valence shells are occupied by four pairs of electrons, the so-called 'rule of eight'. When we come to consider the non-transition elements of the second row of the Periodic Table we find compounds which can only be explained if the central atom has ten or sometimes twelve electrons in its valence shell. Phosphorus pentafluoride and sulphur hexa-fluoride are examples of this type of compound. Phosphorus has five valence electrons and sulphur six. Clearly, all these must be used in PF_5 and SF_6. Nitrogen and oxygen have the same number of valence electrons, and we must ask why phosphorus and sulphur can expand their valence shell to ten electrons while nitrogen and oxygen cannot. The reason is that in the second-row elements the d orbitals can be brought into use. The $3d$ orbitals are in fact of slightly higher energy than the $4s$ orbitals but they are of approximately the same size as the $3s$ and the $3p$ orbitals. This means that when we construct hybrid orbitals of the $3s$, p, and d orbitals each component can make an appreciable overlap with the orbitals of an adjacent atom at the same bond distance. If the $3d$ orbitals were much larger than $3p$

orbitals the distance where the $3d$ orbital had a large overlap the $3p$ would only have a small overlap. The excitation energy for an electron in a $2p$ orbital of an oxygen atom to the $3d$ orbital is not much larger than the energy required to promote an electron from a $3p$ orbital to a $3d$ orbital in the sulphur atom. But the $3d$ orbital is much larger than a $2p$ orbital so that any $3d$ orbital character added to $2s2p$ hybrids does not significantly improve their bonding potential.

In Parts 1 and 2 we have argued that, in general, electron pair bonds will be arranged in such a way that they are as far apart from each other as possible so as to reduce the repulsion energy between them. On this basis we would expect phosphorus pentafluoride to have a trigonal bipyramid structure, and experiment shows that this is the case.

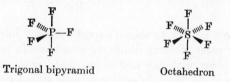

Trigonal bipyramid   Octahedron

It is usual to consider these bonds as sp^3d hybrids constructed from the $3s$ orbital, the three $3p$ orbitals, and the $3d_{z^2}$ orbital (it should be noticed that we could alternatively use the $d_{x^2-y^2}$ and d_{xy} orbitals in place of the p_x and p_y so that we would have an spd^3 hybrid).

Sulphur hexafluoride has an octahedral shape, i.e. the molecule has maximum symmetry with the fluorine atoms placed along the x, y, and z axes to form a regular octahedron. The usual description of the σ bonds in this molecule is based upon the overlap of six $3sp^3d^2$ hybrid wave functions overlapping the fluorine $2p_x$ atomic orbitals. Since PF_5 and SF_6 exist we may ask why ClF_7 does not. The answer is probably that the chlorine atom simply is not large enough to attach seven fluorine atoms around it at a distance which would give strong bonds. A compound IF_7 does exist but the highest valency chlorine fluoride is ClF_3.

d_π–p_π Bonds

A very striking difference between compounds formed by the elements of the second row of the Periodic Table and the compounds formed by the corresponding elements in the first row is the lack of use of p orbitals for π bonding. The second-row elements do not in general form double bonds of the type characteristic of carbon, nitrogen, and oxygen. However, they do make considerable use of their d orbitals to form π bonds particularly with oxygen but also with nitrogen and fluorine. Let us consider the case of dimethyl sulphoxide. Dimethyl sulphide $(CH_3)_2S$ has bond angles of nearly 105°, and we could visualize the bonds as being formed from sulphur $3sp$ hybrid orbitals with the remaining sulphur electrons in non-bonding lone pair orbitals. If we now consider the bonding in dimethyl sulphoxide, the sulphur atom has given up a major share of one pair of electrons to the more electronegative oxygen atom.

If this picture is correct we should expect dimethyl sulphoxide to be an extremely polar molecule behaving as a base. In fact it is not a particularly polar molecule and is no more basic than acetone. However, two of the $3d$ orbitals of the sulphur atom have a symmetry such that they can interact in a bonding sense with the $2p_y$ and $2p_z$ orbitals of the oxygen atom. Thus, in addition to a strong σ bond we would have two extra bonds of the π type. These supplementary π bonds are not very strong, for the electron pairs still belong mostly to the oxygen atom, but this back donation from the oxygen atom helps to counterbalance the polarity of the σ bond. The π bonds add enough strength to bring the total sulphur–oxygen bond energy to about 525 kJ (i.e. 125 kcal). This is still about 25% less than the carbon–oxygen bond in formaldehyde. On the basis of bond strength and bond distance the sulphur–oxygen bond has slightly less double-bond character than a carbonyl bond but electronically it is a triple bond.

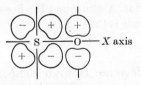

Sulphur $3d_{zx}$ orbital and oxygen $2p_z$ orbital overlapping to produce a d_π–p_π bond

We can visualize the attachment of a further oxygen atom using the remaining sulphur lone pair of electrons in dimethyl sulphoxide; this would give us dimethyl sulphone, $(CH_3)_2SO_2$. According to our octet picture the sulphur atom would now carry a double positive charge and each oxygen atom a negative charge.

$$H_3C \diagdown \overset{++}{\underset{H_3C \diagup}{S}} \diagup \overset{\cdot\cdot}{\underset{\cdot\cdot}{\overset{\cdot\cdot}{O}}} ^{-}$$

If we visualize the four σ bonds as being formed from the usual $3sp^3$ hybrid orbitals in the sulphur there are no single $3d$ orbitals which will make good π bonds with either oxygen atom. There are, however, a variety of $3d$ combinations which will serve just as well. In more sophisticated calculations $4s$ or $4p$ orbitals may also be incorporated. Evidence from bond lengths and bond angles supports the idea that there is considerable d_π–p_π character in the sulphur–oxygen bonds in this type of molecule.

We must now return to sulphur hexafluoride. If we regard this molecule as formed from six $3sp^3d^2$ hybrid orbitals, then the $3d$ functions not used in σ bonding are just those which are ideally suited to π bonding. By mixing these functions with three $4p$ functions on a $\frac{1}{2}p\frac{1}{2}d$ basis we obtain six hybrids with π symmetry extending towards the $2p_y$ and $2p_z$ orbitals of each fluorine atom. Thus each fluorine atom can make two identical π bonds in mutually perpendicular planes. But, since there are only six π bonding functions and six fluorine atoms, we can write structures in which some fluorine atoms have two π bonds and others none. The total resonance hybrid of all such interactions would imply a very strongly bonded structure, and experimentally sulphur hexafluoride is one of the stablest and least reactive of all known molecules. An additional reason for its stability is probably associated with the fact that the fluorine atoms are so tightly packed around the sulphur atom that it is impossible for any attacking species to reach the sulphur atom.

The importance of d_π–p_π bonding in the compounds of phosphorus and sulphur is still very much a matter of controversy. There is no doubt however that the concept of interaction of this kind greatly helps to explain some of the chemical properties of the organic compounds containing phosphorus and sulphur.

The representation of bonds involving $d_\pi-p_\pi$ bonding is a matter of some difficulty. Usually the bonds between sulphur or phosphorus and oxygen are drawn as double bonds with two lines joining the atoms identical to the two lines joining the two carbon atoms in ethylene. Really this is a very incorrect thing to do. Firstly, the bonds are really 'triple bonds' involving six electrons; secondly, it implies that the bonds are similar to those in a carbonyl compound, say, and finally it can lead us into all sorts of errors when we start drawing curved arrows to and from atoms involved in $d_\pi-p_\pi$ bonding. In this book we are therefore going to use 'broken triple bonds' to represent the back donation of electrons from oxygen atoms to sulphur or phosphorus in $d_\pi-p_\pi$ bonds. Thus dimethyl sulphoxide and dimethyl sulphone are represented as depicted below.

Formulae for dimethyl sulphoxide and dimethyl sulphone with 'broken' triple bonds representing the back-donation of electrons from the oxygen atom via $d_\pi-p_\pi$ bonds

We must emphasize that this convention is original to this book and will not be found in the literature elsewhere at the present time, although we hope that other authors will appreciate its merits.

^{31}P Nuclear Magnetic Resonance

The ^{31}P nucleus, like the proton, has a spin of $\frac{1}{2}$, and NMR spectra are important in phosphorus chemistry. Chemical shifts (usually measured relative to 85% phosphoric acid) are usually larger than in proton magnetic resonance [e.g. PH_3 $\delta = +238$ Hz and PCl_3 $\delta = -220$ Hz]. Tetrahedral compounds containing the phosphoryl group are effectively more shielded and their chemical shifts are close to that of the standard [e.g. PO_4^{3-} $\delta = -5$ Hz; $(C_2H_5O)_3PO$ $\delta = +1$ Hz].

The Chemistry of the Phosphines and Dialkyl Sulphides (Thioethers)

The greatest difference between trivalent nitrogen and bivalent oxygen compounds on the one hand and the corresponding trivalent phosphorus and bivalent sulphur compounds on the other is the ability of the latter atoms to increase their covalency during reactions.

Oxygen or oxidizing agents are particularly effective in this respect. Phosphines, unless pure, are spontaneously inflammable in the air, and in all circumstances they are very readily oxidized. Dialkyl sulphides are less reactive but can be oxidized to sulphoxides and ultimately to sulphones.

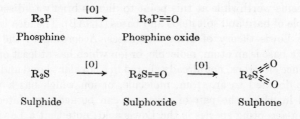

$$R_3P \xrightarrow{[O]} R_3P{=\!\!=}O$$

Phosphine   Phosphine oxide

$$R_2S \xrightarrow{[O]} R_2S{=\!\!=}O \xrightarrow{[O]} R_2S{\overset{\displaystyle O}{\underset{\displaystyle O}{\lessgtr}}}$$

Sulphide   Sulphoxide   Sulphone

Notice that phosphines can only take on one oxygen atom to yield phosphine oxides whereas sulphides can take on two oxygen atoms successively and yield first sulphoxides and then sulphones.

In sharp contrast to ammonia, phosphine itself shows practically no basic properties and the crystalline phosphonium salt $(PH_4^+I^-)$ is hydrolysed instantly by water. The situation is very different when the hydrogen atoms in phosphine are replaced by alkyl groups, and trimethylphosphine is a relatively strong base though less basic than trimethylamine. Notwithstanding the fact that trimethylamine is a slightly stronger base than trimethylphosphine, trialkylphosphines are in general more powerful nucleophiles than the corresponding amines. Thus, (dimethylamino)dimethylphosphine reacts with methyl iodide to give a quaternary salt in which the additional methyl group is attached to the phosphorus, not the nitrogen atom.

$$\underset{H_3C}{\overset{H_3C}{{>}}}\!P\!-\!N\!\underset{CH_3}{\overset{CH_3}{{<}}} \xrightarrow{CH_3I} \overset{H_3C}{\underset{H_3C}{H_3C\!-\!\overset{+}{P}\!-\!\overset{..}{N}}}\!\underset{CH_3}{\overset{CH_3}{{<}}} \quad I^-$$

Just as phosphine is much less basic than ammonia, so hydrogen sulphide is much more acidic than water; similarly mercaptans are more acidic than the corresponding alcohols. Also paralleling the behaviour of phosphines, sulphides are more powerful nucleophiles than ethers and will react slowly with alkyl halides at room temperature to yield the corresponding sulphonium salts.

$$\underset{C_2H_5}{\overset{C_2H_5}{>}}S: \ + \ CH_3I \ \longrightarrow \ \underset{C_2H_5}{\overset{C_2H_5}{>}}CH_3{-\!\!}S:^+ \ I^- \ \xrightarrow{\ Ag_2O\ } \ \underset{C_2H_5}{\overset{C_2H_5}{>}}CH_3{-\!\!}S:^+ \ OH^-$$

Notice that both phosphines and sulphides are less basic but more nucleophilic than the corresponding amines and ethers.

It seems worthwhile at this point to digress briefly to discuss the principle of hard and soft acids and bases (HSAB). The idea is based on the Lewis theory of acids and bases. According to the Lewis theory a base is an atom, molecule, or ion which has at least one pair of valence electrons, not already shared in a covalent bond and which can be donated to an atom, molecule, or ion which has a vacant orbital in which the pair of electrons can be accommodated. The electron-accepting species is the Lewis acid; note that a Lewis base is identical with the conventional Brønsted–Lowry base. We can depict an acid–base reaction in Lewis theory as follows:

$$A + B: \ \rightarrow \ A:B$$

The species A:B may be called an acid–base complex. Hydrogen chloride, a Brønsted–Lowry acid, becomes an acid–base complex in the Lewis picture, in which the proton H^+ is the Lewis acid and the chloride anion Cl^- is the base. It is not our present purpose to discuss the Lewis theory of acids and bases in any detail. The concept can be extended and ethyl alcohol can for example be regarded as being composed of the ethyl cation $C_2H_5^+$ and the hydroxide anion OH^-. Methyl chloride can be regarded as being an acid–base complex made up of the methyl cation CH_3^+ and the chloride anion Cl^-, while methane can be regarded as either a combination of CH_3^+ and H^- or more usually H^+ and CH_3^-.

Let us now return to consider a series of Lewis acids and bases which can exist in equilibrium. In a qualitative way the strength of an acid is a measure of the stability of the bond it will form in the particular complex B. Thus, if acid A′ displaces acid A from the complex

$$A' + A{:}B \ \rightarrow \ A'{:}B + A$$

A:B then we should regard A′ as being the stronger acid. Conversely if base B′ displaces base B from the complex then we would regard B′ as the stronger base

$$:B' + A{:}B \ \rightarrow \ A{:}B' + :B$$

If we now compare a series of acids and bases of this type we find that there are cases where base B' is stronger than base B when acid A is the complexing acid, but that the situation is reversed when acid A' is the complexing acid. This absence of a universal scale of acid and base strength is summarized in the qualitative and empirical rule called the Principle of Hard and Soft Acids and Bases. The rule states that *hard acids prefer to bind with hard bases and soft acids prefer to bind with soft bases*. A hard base is one in which the donor atom is of high electronegativity, low polarizability, and is hard to oxidize. A soft base is one in which the donor atom is of low electronegativity, high polarizability, and is easy to oxidize. A hard acid is one in which the acceptor atom is of small size, of high electronegativity, and of low polarizability, whereas a soft acid has a large acceptor atom of high polarizability and comparatively low electronegativity. The Lewis acids and bases common in organic chemistry are listed in Table 5.2.

Table 5.2. Hard and soft acids and bases

	Hard	Borderline	Soft
Bases			
	F^-	$C_6H_5NH_2$	I^-
	OH^-	C_5H_5N	RS^-
	$CH_3CO_2^-$	N_3^-	SCN^-
	RO^-	Br^-	CN^-
	H_2O	NO_2^-	R^-
	ROH	SO_3^{2-}	RSH
	NH_3		R_2S
			R_3P
Acids			
	H^+	NO^+	Hg^{2+}
	Li^+	R_3C^+	I^+
	Na^+		Br^+
	BF_3		Quinones
	$AlCl_3$		Carbenes
	RCO^+		$(CN)_2C{=}C(CN)_2$

It is not our purpose to discuss the principle of hard and soft acids and bases any further in the present chapter. The concept is probably more useful in inorganic chemistry than in organic chemistry. We just wish to note that the observations we have made about the relative nucleophilicity of phosphines and sulphides on the one hand, and amines and ethers on the other is completely in accord with the concept. The methyl cation CH_3^+ is a moderately soft acid centre and hence soft bases such as RS^- or R_3P react very rapidly with methyl chloride, CH_3Cl. On the other hand the carbonyl group behaves as a hard acid centre (CH_3CO^+) and is attacked by hard bases (OH^-, OR^-, H_2O, NH_3), but it is not significantly attacked by soft bases (I^-, RS^-). The principle of hard and soft acids and bases describes a wide range of chemical phenomena in a qualitative but not a quantitative way. It is not infallible and many apparent discrepancies exist. Its use is that it helps to correlate a large amount of data and it can be used tentatively for prediction. We shall now return to our discussion of the chemistry of phosphines and sulphides.

Tertiary phosphines are most readily prepared from phosphorus tribromide and the corresponding Grignard reagent.

$$PBr_3 + 3C_2H_5MgBr \rightarrow (C_2H_5)_3P + 3MgBr_2$$

We have already emphasized that there are two great differences from amines; first, the ease with which the phosphorus atom assumes a pentavalent state, particularly as witnessed by the ease of oxidation, and secondly, the greater nucleophilicity of the phosphine. Optically active phosphines can be prepared. The inversion frequency of tertiary amines is so high that optical isomers have not yet been obtained; the inversion frequencies of trivalent phosphorus compounds, on the other hand, are low so that optical isomers stable at room temperatures have been isolated. This was achieved by the electrolytic reduction of optically active phosphonium salts.

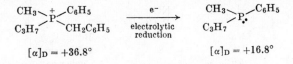

The preparation of thiols and sulphides by the displacement reaction of an alkyl halide was discussed briefly in Chapter 3 of Part 1. In general, the chemistry of thiols and sulphides is much closer to the chemistry of alcohols and ethers than is the chemistry of phosphine related to the chemistry of amines. Sulphides can be oxidized to sulphoxides and eventually to sulphones but the reaction occurs far less readily than the corresponding oxidation of phosphines to phosphine oxides. With mild oxidizing agents mercaptans are oxidized to disulphides. This is a reaction in which they differ markedly from alcohols where the formation of peroxides does not occur readily.

$$2 \text{ RSH} \xrightarrow{\text{[O]}} \text{R–S–S–R}$$

The oxidizing agent may be Cl_2, HNO_3, H_2SO_4, $KMnO_4$, etc., or O_2 in alkaline solution. Disulphides are much more stable than peroxides but if they are heated sufficiently they undergo homolytic fission to thiyl radicals, analogous to the splitting of peroxides yielding alkoxide radicals.

Phosphonium and Sulphonium Ylids

The term 'ylid' was first coined by the German chemist Wittig and is now used to describe the compound in which a carbanion is attached directly to a heteroatom carrying a positive charge. Some authors translate into English and write 'ylide'. The two most important examples are phosphonium ylids and sulphonium ylids.

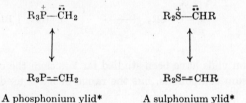

A phosphonium ylid*   A sulphonium ylid*

One can look at these compounds in terms of the 'hard and soft acids and bases' concept. Carbene, CH_2, is a Lewis acid; it has only six valence electrons but it is a softer acid than the methyl cation, CH_3^+, since the latter has an additional proton which will have a 'hardening' effect. Thus ylids such as $CH_2{=}{=}SR_2$ or $CH_2{=}{=}PR_3$ will be more stable than the corresponding oxygen or nitrogen ylids

* We have drawn phosphorus–oxygen and sulphur–oxygen bonds involving d_π–p_π bonding as broken triple bonds. The carbon–phosphorus or carbon–sulphur bond in the ylids can only involve one p atomic orbital of carbon, hence we draw a broken double bond ($R_3P{=}{=}CH_2$, $R_2S{=}{=}CHR$).

$\overset{=}{C}H_2-\overset{+}{O}R_2$ or $\overset{=}{C}H_2-\overset{+}{N}R_3$. The most widely applicable method for preparing phosphonium ylids is the treatment of the corresponding quaternary phosphonium salt with a suitable base.

$$\text{B:} \quad \overset{\displaystyle H}{\underset{\displaystyle R''}{R'-C-\overset{+}{P}(C_6H_5)_3}} \longrightarrow \overset{R'}{\underset{R''}{>}}C{=}P(C_6H_5)_3 + BH^+$$

A wide variety of different bases have been used in the formation of phosphonium ylids. Carbanion bases such as tritylsodium, butyl-lithium, and phenyllithium have been used extensively. Sodamide has also been used, and for acidic phosphonium salts ammonia, sodium hydroxide, and even sodium carbonate have been used. When alkyllithium compounds are used as bases there is a possibility of ligand exchange, a risk which appears to be reduced but not eliminated by using phenyllithium.

Sulphonium ylids can be formed in a similar manner, by treating sulphonium salts containing an α-hydrogen atom with an appropriate base.

$$\text{B:} \quad \overset{\displaystyle H}{\underset{\displaystyle R''}{R'-C-\overset{+}{S}R_2}} \longrightarrow \overset{R'}{\underset{R''}{>}}C{=}SR_2 + BH^+$$

Sulphonium ylids have been studied far less than the corresponding phosphonium compounds, but the range of bases used is much the same.

Both phosphonium and sulphonium ylids behave as dipolar molecules containing stabilized carbanions, and their most important chemical reactions are those characteristic of carbanions. The reactions of phosphonium ylids have been very extensively studied but the only reaction we are going to discuss is that of the ylids with carbonyl compounds. Phosphonium ylids will react with aldehydes or ketones to form an olefin and a phosphine oxide. This is known as the Wittig reaction.

$$R_3P{=}CR_2' + R_2''C{=}O \rightarrow R_2''C{=}CR_2' + R_3PO$$

$$\text{Wittig reaction}$$

Sulphonium ylids, on the other hand, react with aldehydes and ketones to yield ethylene oxides and sulphides.

$$R_2S\!=\!\!=\!CR'_2 + R''_2C\!=\!\!=\!O \longrightarrow R''_2C\!\!-\!\!CR'_2 + R_2S$$
$$\diagdown\!\!O\!\!\diagup$$

Both of these reaction sequences have considerable application in synthesis, and the Wittig reaction in particular has been extensively used in organic synthesis.

The Wittig reaction is believed to involve an activated complex in which the valency of four atoms is changing in a kind of electrocyclic reaction.

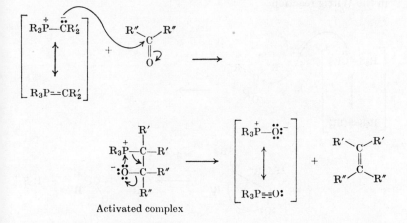

Activated complex

The reaction is quite general and works with aldehydes as well as with ketones. It also works with unsaturated carbonyl compounds such as ketenes and with conjugated ketones. (In the latter case the nucleophilic centre of the Wittig reagent usually attacks the carbonyl group, rather than undergoing conjugate addition.) In many cases the Wittig reagent may be prepared *in situ* so that the solution containing the carbonyl compound and the required phosphonium salt may be treated directly with a base.

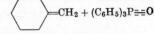

(Wittig reagent prepared *in situ.*

Base = C_6H_5Li, yield 48%
Base = Bu^tOK, yield 91%)

The Wittig reaction has been used extensively in natural product syntheses including the preparation of all-*trans*-squalene, β-carotene, vitamin A, and vitamin D (see Part 4).

The reaction of sulphonium ylids with carbonyl compounds has been much less extensively studied, although it appears to be a quite general method of preparing ethylene oxides. Again the reaction is initiated by the nucleophilic addition of the ylid to the carbonyl compound. The activated complex in this case involves a 3-membered electrocyclic reaction in place of the 4-membered process involved in the Wittig reaction.

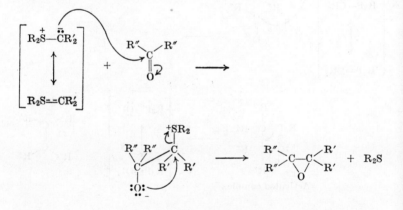

We immediately ask the question, why is there a difference between the phosphonium and the sulphonium ylids? There is no certain answer although we may note that a phosphorus atom in a trialkyl-phosphine is more readily oxidized to the phosphine oxide than the sulphur atom in a sulphide. Closely connected is the fact that a sulphide is an excellent leaving group. The reaction, as we have depicted it, involves the antiperiplanar displacement of the sulphur atom by the oxygen atom. This is in accord with the observation that the reaction between benzaldehyde and diphenylsulphonium benzylid yields exclusively the *trans*-stilbene oxide.

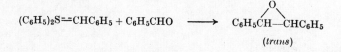

Notice that in the *erythro*-intermediate required for the *trans*-stilbene oxide the bulky groups are to some extent staggered whereas in the *threo*-intermediate which would lead to the *cis*-ethylene oxide all the bulky groups are concentrated together.

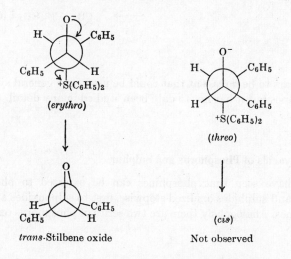

A number of other sulphur ylids can be prepared but we only have space to mention the oxosulphonium ylid prepared from dimethyl sulphoxide via the trimethyloxosulphonium iodide which on treatment with sodium hydride yields the oxosulphonium ylid.

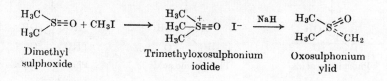

The reactions of this ylid are analogous to those of the sulphonium ylids we have described above.

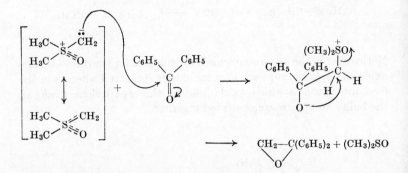

It appears to be a reagent that could be used for a general synthesis of oxiranes but it has not to date been studied in any detail.

The Oxyacids of Phosphorus and Sulphur

We have seen that phosphines can be oxidized to phosphine oxides and sulphides oxidized stepwise, first to sulphoxides and then sulphones. Analogously there are two series of phosphorus oxyacids:

$$\begin{array}{ccc}
\underset{R}{\overset{R}{>}}P-OH & \underset{R}{\overset{HO}{>}}P-OH & \underset{HO}{\overset{HO}{>}}P-OH \\
\text{Phosphinous acids} & \text{Phosphonous acids} & \text{Phosphorous acid} \\
\downarrow [O] & \downarrow [O] & \downarrow [O] \\
\underset{R}{\overset{R}{>}}\underset{O}{\overset{OH}{P}} & \underset{HO}{\overset{R}{>}}\underset{O}{\overset{OH}{P}} & \underset{HO}{\overset{HO}{>}}\underset{O}{\overset{OH}{P}} \\
\text{Phosphinic acids} & \text{Phosphonic acids} & \text{Phosphoric acid}
\end{array}$$

and three series of sulphur oxyacids:

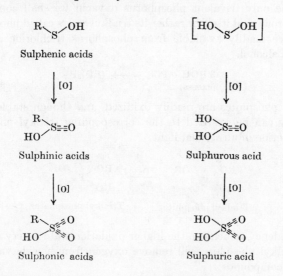

Sulphenic acids

Sulphinic acids

Sulphurous acid

Sulphonic acids

Sulphuric acid

We must further note that the trivalent phosphorus acids are capable of tautomerism as are both the bivalent and the trivalent sulphur oxyacids.

Tautomerism possible in phosphinous and phosphonous acids

Tautomerism possible in sulphenic and sulphinic acids

In practice the oxyacids of phosphorus and sulphur in the not fully oxidized state are unstable substances and, although they show some interesting chemistry, this is outside the scope of our present discussion. The only trivalent phosphorus oxyacid we shall consider is phosphorous acid itself or rather its trialkyl esters called phosphites. These are readily available from phosphorus trichloride and the required alcohol.

$$3 \text{ ROH} + PCl_3 \xrightarrow{\text{Base}} (RO)_3P$$
(excess)

Trialkyl phosphites are readily oxidized, and though stable in the air, they can be oxidized to the corresponding trialkyl phosphate by the action of ultraviolet light.

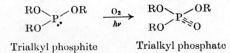

Trialkyl phosphite   Trialkyl phosphate

This tendency to achieve the higher oxidation state is very marked, and trialkyl phosphite will remove oxygen from a wide variety of organic compounds.

$$2 \text{ ArCHO} \xrightarrow[160° \text{ (yields poor)}]{(EtO)_3P} ArCH{=}CHAr + 2(EtO)_3P{=\!=}O$$

$$Ph_2C{=}C{=}O \xrightarrow[\text{Heat}]{(EtO)_3P} PhC{\equiv}CPh + (EtO)_3P{=\!=}O$$

$$PhN{=}C{=}O \xrightarrow{(EtO)_3P} PhN{\equiv}C + (EtO)_3P{=\!=}O$$

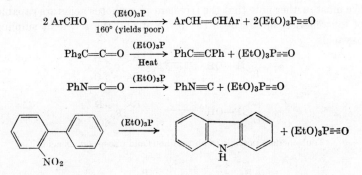

It appears that the first two reactions might involve a carbene, and the fourth reaction a nitrene. The important point is that the desire of the phosphorus atom to attain the higher valency state provides a driving force for these reactions. In all these examples the phosphorus atom has obtained the phosphoryl oxygen atom from another molecule but in some reactions this oxygen is acquired as the result of rearrangement which we can illustrate by the important Arbusov reaction. A trialkyl phosphite reacts with an alkyl halide to form the corresponding phosphonium salt as one would expect, but this salt rearranges to give the new alkyl halide and a dialkyl alkylphosphonate. This reaction provides a simple synthesis of phosphonic diesters.

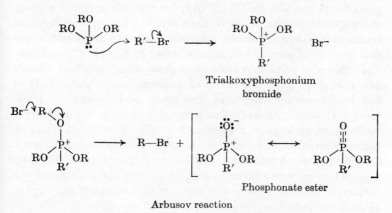

Trialkoxyphosphonium bromide

Phosphonate ester

Arbusov reaction

With α-halogeno-ketones a rather interesting variation of this reaction occurs. The phosphite adds to the carbonyl bond but then transfers to the oxygen atom with concomitant ejection of a halide ion.

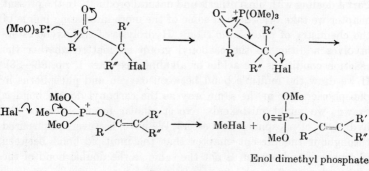

Enol dimethyl phosphate

Perkov reaction

The halide ion then attacks one of the original methyl groups of the phosphite in a reaction exactly analogous to the second stage of the

Arbusov reaction. The overall result is to give an enol phosphate ester, and the reaction is sometimes known as the Perkov reaction.

Phosphonic acid derivatives have received considerable study on account of their applications in industry and medicine. Isobutane-1-phosphonyl fluoride, $PrO(Me)P(O)F$, for example is one of a group of compounds known as nerve gases and is lethal in very low concentrations indeed. Aromatic sulphonic acids are important in the manufacture of phenols and naphthols (see Part 2, pp. 215 and 264). The amides of sulphonic acids have great pharmaceutical interest (see Part 1, p. 101). The esters of sulphonic acids, particularly of toluene-p-sulphonic acid, have been much used, because, on treatment with alkali, instead of undergoing hydrolysis like carboxylic esters, they undergo S_N2 type nucleophilic substitution. Thus an optically active alcohol can be converted into its toluene-p-sulphonyl ester by treatment with toluene-p-sulphonyl chloride, and when this ester is treated with sodium hydroxide (or with the sodium salt of a carboxylic acid) it is converted into the enantiomer of the original alcohol (or its carboxylic ester) (see Part 2, pp. 334—5). This type of reaction has been particularly important in the chemistry of sugars (see Chapter 2 of Part 4).

The most important organic phosphorus derivatives of all are the esters of phosphoric acid. These esters, particularly esters of sugars, occur widely in all living organisms and they play a vital role in many biological processes. These will be discussed in the chapters of Part 4 dealing with biosynthesis and natural products. In the present chapter we take brief note of some of the more interesting points of the chemistry of phosphoric esters. Hydrolysis of carboxylic esters involves addition to the carbonyl group no matter whether the reaction conditions are acidic or alkaline (see Part 1, pp. 86—88). If we drew the multiple bond between oxygen and phosphorus in phosphoric acids in the same way as the carbonyl double bond is drawn, we might mistakenly expect similar mechanisms for the hydrolysis of phosphoric acid esters. We have, however, emphasized throughout the present chapter that the multiple bond between phosphorus and oxygen is not the same as the double bond of the carbonyl group.*

---

* This is not to say that pentacovalent phosphorus is *never* involved as an intermediate, but simply that the great majority of nucleophilic reactions do not involve opening of the phosphoryl bond.

There are two well-defined mechanisms for the hydrolysis of trialkyl or triaryl phosphates, and in neither case is there evidence for addition to the phosphoryl bond. In alkaline hydrolysis in water enriched with ^{18}O, the oxygen isotope is not incorporated into the ester before hydrolysis, i.e. nucleophilic addition does *not* occur.

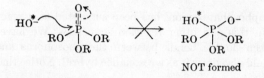

NOT formed

Instead nucleophilic displacement takes place, analogous to the S_N2

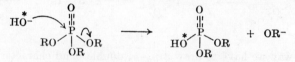

Alkaline hydrolysis of a trialkyl phosphate

reactions of alkyl halides. In acidic media no ^{18}O is incorporated into either the starting ester or the resultant acid; instead the label appears in the alcohol. The reaction therefore involves alkyl–oxygen fission.

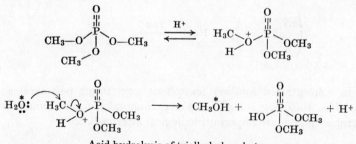

Acid hydrolysis of trialkyl phosphate

The neutral or alkaline hydrolysis of phosphoric mono- or di-esters involves a different situation because the ester may itself be ionized. The rate of hydrolysis of phosphoric acid monoesters varies with pH, reaching a maximum around pH 4. The reaction in this case involves phosphorus–oxygen fission and is probably a unimolecular process.

The 'metaphosphate anion' has been suggested as a reactive intermediate in this and other reaction sequences. We have drawn it with broken double bonds between the phosphorus and oxygen, but we could also draw it as a resonance hybrid. Notice that depicted

in this way we have to draw a $p_\pi - p_\pi$ double bond characteristic of the first row of the Periodic Table. Evidence for the existence of the metaphosphate ion is not yet conclusive, although some very reactive intermediate must be involved in these reactions. The 'metaphosphate ion' reacts extremely rapidly not only with water but with alcohols, and is in fact a powerful phosphorylating agent.

In concentrated solution phosphoric acid forms polymeric anhydrides, the dimer being known as pyrophosphoric acid. Monoalkyl pyrophosphates are important biological intermediates.

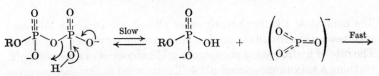

Pyrophosphate ester   Phosphate ester   'Metaphosphate ion'

Triphosphates are also important in biological systems. Both pyrophosphates and triphosphates function as phosphorylating agents, probably via the unimolecular elimination of the metaphosphate ion. The triphosphates are more reactive than the pyrophosphates in this respect.

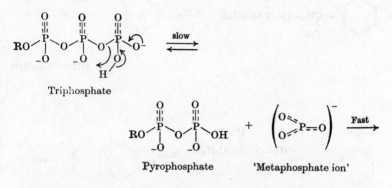

Triphosphate

Pyrophosphate   'Metaphosphate ion'

We described above how acidic hydrolysis of a trialkyl phosphate involves alkyl–oxygen fission. This implies that phosphoric acid derivatives are good leaving groups. This can be illustrated by the reaction of aniline with trimethyl phosphate which gives dimethylaniline in an exothermic process. Whether a nucleophile attacks carbon or phosphorus depends on the nature of the substituents and also on the 'hardness' of the nucleophile, e.g.

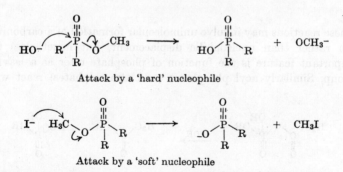

Attack by a 'hard' nucleophile

Attack by a 'soft' nucleophile

Carbanions are borderline bases, and when attack on carbon occurs with the phosphate acting as a leaving group we have a method

9

of lengthening carbon chains which is believed to occur extensively in biological processes. Laboratory experiments of this kind include:

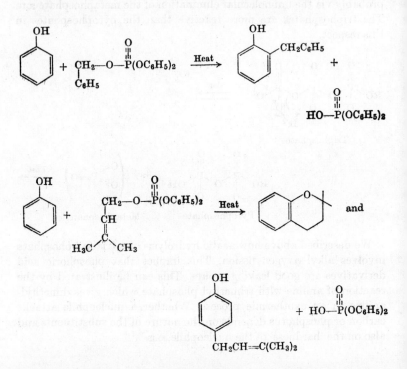

These reactions may involve unimolecular formation of a carbonium ion rather than nucleophilic displacement, but nonetheless the important feature is the function of phosphate ester as a leaving group. Similarly acyl phosphates (and also sulphates) react with

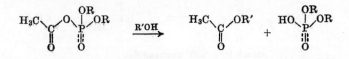

alcohols to give acyl esters. The 'leaving group' behaviour of phosphate is also important in elimination reactions.

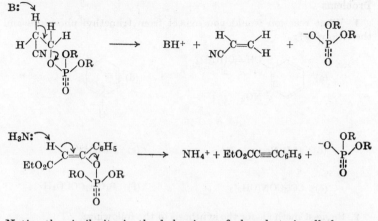

Notice the similarity in the behaviour of phosphate in all these reactions to the behaviour of halogen in many synthetic processes discussed in Parts 1 and 2. We shall see in later chapters in Part 4 that in biological systems phosphate takes the role of halide, and where in the laboratory we would synthesize a halogeno-derivative prior to a nucleophilic displacement or elimination, in nature exactly the same sequences involve prior formation of a phosphate derivative.

Bibliography

General

A. J. Kirby and S. W. Warren, *The Organic Chemistry of Phosphorus*, Elsevier, Amsterdam/London/New York, 1967.

R. F. Hudson, *Structure and Mechanism in Organo-Phosphorus Chemistry*, Academic Press, London, 1965.

[These two books complement each other. Kirby and Warren's book provides an excellent account of the chemistry of phosphoric acid derivatives. Hudson's book gives a better coverage of trivalent phosphorus and theory in general.]

Special topics

A. William Johnson, *Ylid Chemistry*, Academic Press, New York and London, 1967.

S. Trippett, "Wittig Reaction," Chapter 3 in *Advances in Organic Chemistry*, Vol. I, Interscience, New York, 1960.

J. I. G. Cadogan, "Reactions involving Tervalent Phosphorus Compounds", in *Quarterly Reviews*, 1962, p. 208; 1968, p. 222.

Problems

1. What reaction would you expect from trimethyl phosphite and the following?:

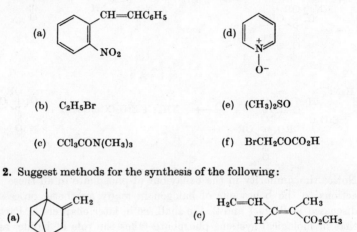

(a)

$$CH=CHC_6H_5$$
$$NO_2$$

(d)

(b) C_2H_5Br

(e) $(CH_3)_2SO$

(c) $CCl_3CON(CH_3)_3$

(f) $BrCH_2COCO_2H$

2. Suggest methods for the synthesis of the following:

(a)

$$CH_2$$

(c)

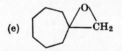

$$H_2C=CH$$
$$H$$
$$C=C$$
$$CH_3$$
$$CO_2CH_3$$

(b) $C_6H_5CH=CH—CH=CHC_6H_5$

(d) $C_6H_5COCH=C(C_2H_5)_2$

(e)

$$CH_2$$

3. Predict the outcome of the following reactions:

$$(C_6H_5O)_2P—O—P(OCH_2C_6H_5)_2 +$$ (a) NH_3

(b) I^-

(c) OH^-

CHAPTER 6

Physical Organic Chemistry and the Establishment of Reaction Mechanisms

Throughout Parts 1 and 2 the chemical reactions of carbon compounds have been discussed on a mechanistic basis. The greatest justification for all the mechanistic schemes presented is that they form a coherent whole. In spite of the wide variety of the carbon compounds described, the number of basic mechanisms invoked is comparatively few, but nevertheless they provide a logical scheme into which the experimental facts of organic chemistry can be fitted. This alone would be more than ample justification for the mechanistic schemes developed. In a sense, in Parts 1 and 2 we have put the cart before the horse insofar as the very successful mechanistic theories of organic chemistry today would not have been developed if it had not been for physico-chemical studies that preceded the theoretical interpretation.

The mechanism of a reaction is literally the detailed pathway by which the reaction occurs, and a knowledge of it involves much more than simply a knowledge of the nature of the system before and after reaction. Consider, for example, the addition of hydrogen bromide to ethylene. The chemical equation gives us little information about the mechanism of the reaction.

$$HBr + CH_2{=}CH_2 \rightarrow CH_3CH_2Br$$

For instance it does not tell us whether the hydrogen and bromine atoms become attached to their carbon atoms simultaneously or in steps. Again, if the second possibility is true, it does not tell us whether the hydrogen or the bromine atom becomes attached first. When we can answer questions such as these, we are describing the reaction mechanism.

The Study of Reaction Rates

Before proposing the mechanism of a reaction it would be a great help if the actual number of molecules or ions involved in the rate-determining process could be known. This is known as the *molecularity*. In fact, this is a somewhat indefinite concept since in complicated processes occurring in solution it is often difficult to decide exactly the number of molecules, particularly solvent molecules, taking part in the rate-determining step. Nonetheless, molecularity is a useful concept; but it is important to distinguish it from the *reaction order* which refers to an experimentally measured quantity. The reaction order is the sum of the powers to which the concentration terms, appearing in the equation for the rate of the reaction, are raised.

The simplest type of reaction is one in which A is converted into products. If no other species are involved the rate of this reaction is dependent only on the concentration of A.

$$A \rightarrow Products$$

$$\frac{-d[A]}{dt} = \frac{+d[Products]}{dt} = k[A]^*$$

(Rate of disappearance of A = Rate of formation of products)

k, measured in (seconds)$^{-1}$, is called the first-order rate constant.

A reaction of this type is known as a first-order reaction because the observed rate depends on the concentration of A to the first power. While the rate law given is a probable one for the conversion of A into products, it is by no means the only possibility. Suppose two molecules of A must collide with each other to yield products; the rate law then would be

$$\frac{-d[A]}{dt} = k[A]^2$$

So the order of the reaction must be considered. The rate of the reaction just shown depends on the second power of the concentration

* Dimensions of k are mole litre^{-1} sec^{-1} = k mole litre^{-1}. Therefore the units of a first-order rate constant are sec^{-1}.

of A and is of the second order. A slightly more complicated reaction would be one in which A reacts with B to give products.

$$A + B \rightarrow \text{Products}$$

$$\frac{-d[A]}{dt} = \frac{-d[B]}{dt} = \frac{+d[\text{Products}]}{dt} = k[A][B]*$$

(Rate of disappearance of A =

Rate of disappearance of B = Rate of formation of products)

k, measured in (mole)$^{-1}$ (litres) (seconds)$^{-1}$, is called the second-order rate constant.

Similarly, a reaction whose rate depends on the concentrations of three components A, B, and C is called a third-order reaction, and a reaction whose rate is independent of the concentration of the reacting species is called a zero-order reaction.

If a reaction is not one step it is either a chain reaction (see section on chlorination, p. 312) or a sequence. If it is a sequence the rate is the rate of the slowest step. Complications arise if the early steps are significantly reversible or if the concentrations of certain reagents are constant. For example, the rate of the oxidation of isopropyl alcohol by potassium dichromate in concentrated acid solutions, viz.:

$$3(CH_3)_2CHOH + 2HCrO_4^- + 8H^+ = 3(CH_3)_2CO + 2Cr^{3+} + 8H_2O$$

is found to be proportional to the concentration of the alcohol, the concentration of the dichromate ion, and the square of the acid concentration, i.e.

$$\text{Rate} = k[(CH_3)_2CHOH][HCrO_4^-][H^+]^2$$

The reaction is fourth-order. However, extensive investigations have indicated that the mechanism of this reaction involves two steps.

$$H_2\ddot{O}:$$

---

* Dimensions of k are mole litre^{-1} sec^{-1} = k mole^2 litre^{-2}. Therefore the units of a second-order rate constant are mole^{-1} litre sec^{-1}.

The rate-determining step is bimolecular and one of the reactants is the solvent which does not appear in the rate equation because its concentration is effectively constant. We see then that the experimentally determined order can only be explained once the multiplicity of the mechanism is understood.

Notice that we can write the reaction of methane and chlorine in the gaseous state to yield products in just the same way as we write $A + B$ to yield products.

$$CH_4 + Cl_2 \rightarrow CH_3Cl + HCl$$

$$(A) + (B) \rightarrow (Products)$$

Experimentally, however, it is found that the rate of the reaction is not dependent on the product of the two concentrations. This is because, as indeed we know, the reaction is a complex one and it neither shows second-order kinetics nor is it bimolecular. Similarly, the reaction of t-butyl bromide with hydroxide ions in solution to yield products appears to have the same form as our idealized second-order reaction of $A + B$.

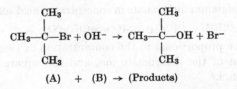

$$(A) \quad + \quad (B) \rightarrow (Products)$$

In practice this reaction does not show second-order kinetics (even under conditions where the concentration of hydroxide ion varies considerably during the course of the reaction), and from this it is deduced that the reaction is probably not bimolecular. Under special conditions, in fact, the reaction rate is found to depend only on the concentration of t-butyl bromide. That is, it shows first-order kinetics.

It should be evident by now to the reader that the observed rate law for a reaction may be of great help in deducing its reaction mechanism. It is obvious, for example, that a reactant which appears in the rate expression must be involved in or before the rate-determining step.

For example, consider the diazotization of aniline in dilute acid:

$$C_6H_5\overset{+}{N}H_3 + HNO_2 \rightarrow C_6H_5\overset{+}{N}\equiv N + 2H_2O$$

The stoichiometry of this reaction demands only one molecule of nitrous acid. However, the reaction has been found to be second-order in nitrous acid:

$$\text{Rate} = k[\text{HNO}_2]^2[\text{C}_6\text{H}_5\text{NH}_2]$$

and this suggests that two molecules of nitrous acid somehow participate in the sequence leading to the rate-determining step.

There is no general way of proceeding from the empirical rate law for a reaction to its mechanism. Usually we list the most plausible mechanisms, derive hypothetical rate laws for each, and then decide which of these corresponds most closely to the observed rate law.

In the above reaction, the kinetic data are consistent with a rate-determining step involving the amine and a species containing two atoms of trivalent nitrogen, probably N_2O_3. The fact that solutions in which diazotizations are performed exhibit frequently the blue colour of N_2O_3 makes this assumption attractive. The sort of mechanism we might therefore propose is as follows.

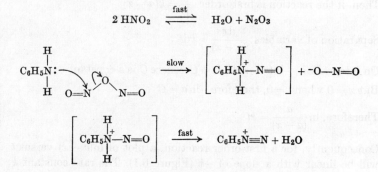

This mechanism is compatible with the observed rate law:

$$\text{Rate} = \text{Rate of slow step}$$

$$= k[\text{C}_6\text{H}_5\text{NH}_2][\text{N}_2\text{O}_3]$$

But

$$\frac{[\text{N}_2\text{O}_3][\text{H}_2\text{O}]}{[\text{HNO}_2]^2} = K$$

where K is the equilibrium constant. Therefore,

$$\text{Rate} = k[C_6H_5NH_2][HNO_2]^2 \frac{K}{[H_2O]}$$

$$= \text{a new constant} \times [C_6H_5NH_2][HNO_2]^2$$

as observed. Since kinetic data can usually furnish no information about the course of a reaction sequence after the rate-determining step, we can say little about the conversion of the N-nitroso-compound into the diazonium cation.

An organic chemist must be able to integrate his rate expression. If we assume at the start of the reaction that the concentration of A is equal to a, and that after time t the concentration of A remaining will be $(a - x)$. That is, in the reaction

$$\text{A} \rightarrow \text{Products}$$

Concentration when $t = 0$: a

Concentration when $t = t$: $a - x$

Then, if the reaction is first-order, $\dfrac{\mathrm{d}x}{\mathrm{d}t} = k(a - x)$

Separation of variables: $\dfrac{\mathrm{d}x}{(a - x)} = k\mathrm{d}t$

On integration: $-\ln(a - x) = kt + C$, where C is a constant.

But $x = 0$ when $t = 0$, therefore $-\ln a = C$

Therefore, $\ln \dfrac{a}{(a - x)} = kt$

Consequently, for a first-order reaction, a plot of $\ln(a - x)$ versus t will be linear with a slope of $-k$ (Figure 6.1). The rate constant k can therefore be evaluated from this slope.

By a similar argument we obtain an analogous expression for a second-order reaction which is bimolecular, e.g.

$$\text{A} + \text{B} \rightarrow \text{Products}$$

Concentration when $t = 0$: a b

Concentration when $t = t$: $a - x$ $b - x$

Therefore, $\dfrac{\mathrm{d}x}{\mathrm{d}t} = k(a - x)(b - x)$

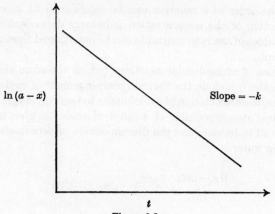

Figure 6.1

Therefore, on integration:

$$\frac{1}{(a-b)} \ln \frac{b(a-x)}{a(b-x)} = kt$$

If A and B are in equal concentration at the start of the reaction, then

$$\frac{\mathrm{d}x}{\mathrm{d}t} = k(a-x)^2$$

which, when integrated, gives

$$\frac{x}{a(a-x)} = kt \qquad \text{(Figure 6.2)}$$

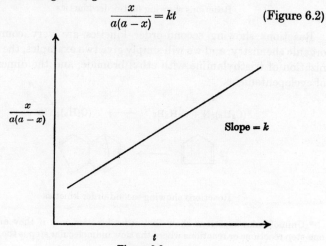

Figure 6.2

Thus the order of a reaction can be determined by ascertaining what function of the concentration is linearly dependent on time, and the values of the rate constants can be determined from the slope of the graph.

Examples of unimolecular reactions (which therefore show first-order kinetics*) include the thermal rearrangement of cyclopropane to propene, the dissociation of cyclobutane to two ethylene molecules, the thermal decomposition of t-butyl chloride to give hydrogen chloride and isobutene, and the decomposition of benzenediazonium chloride in water.

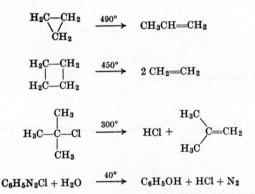

$$C_6H_5N_2Cl + H_2O \xrightarrow{40°} C_6H_5OH + HCl + N_2$$

Reactions showing first-order kinetics

Reactions showing second-order kinetics are very common in organic chemistry, and we will simply give two examples, the quaternization of triethylamine with ethyl bromide, and the dimerization of cyclopentadiene.

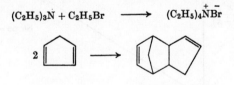

Reactions showing second-order kinetics

* Unimolecular reactions show first-order kinetics only if they are simple one-step reactions or reactions where the slow unimolecular step is the first one.

One of the most familiar general phenomena of chemistry is the increase in the rate of a reaction as the temperature is raised. Empirically, this increase in rate is found to obey the Arrhenius equation

$$k = A\,e^{-E/RT}$$

R is the gas constant, T is the absolute temperature, A is a constant known as the 'A-factor' which contains an entropy term, and E is the activation energy (Part 2, p. 75). By using the transition-state theory, a set of equations relating the rate constant of a reaction to the free energy of activation, the heat of activation, and the entropy of activation can be derived. These are:

$$\Delta G^{\ddagger} = -RT \ln K^{\ddagger} = -RT \ln\left(\frac{kh}{kT}\right)$$

or

$$k = \left(\frac{kT}{h}\right) e^{\Delta G^{*}/RT} \tag{1}$$

and

$$\Delta G^{\ddagger} = \Delta H^{\ddagger} - T\Delta S^{\ddagger} \tag{2}$$

Where k = Boltzmann's constant, T = absolute temperature, and h = Planck's constant.

By substitution in equation (1),

$$k = \left(\frac{kT}{h}\right) e^{\Delta S^{*}/R}\,e^{-\Delta H^{*}/RT} \tag{3}$$

Rearranging and taking logarithms, equation (3) becomes

$$\log\left(\frac{k}{T}\right) = \log\left(\frac{k}{h}\right) + \Delta S^{\ddagger}/2.303R - \Delta H^{\ddagger}/2.303RT$$

and the value of $\Delta H^{\ddagger}$ can be found from the slope of the linear graph of $\log(k/T)$ against $1/T$.

The free energy of activation can be similarly obtained from equation (1), and hence the entropy of activation can be calculated from equation (2).

Typical unimolecular gas-phase reactions, such as the first-order decomposition of t-butyl bromide or nitromethane, or the isomerization of cyclopropane to propene, have positive entropies of activation (see Table 6.1). Qualitatively we can relate the formation of two particles out of one, or the destruction of a cyclic compound, with an increase in randomness and hence an increase in entropy. On

Table 6.1. Unimolecular gas-phase reactions

Decomposition of	$\Delta S^{\ddagger}$ (J °K^{-1} mole^{-1})
t-Butyl bromide	8.4
Nitromethane	8.4
Acetic anhydride	-16.7
Allyl vinyl ether	-20.9
Cyclopropane	46.0

the other hand the decomposition of acetic anhydride, or the isomerization of allyl vinyl ether (see below), have negative entropies of activation. Here we can relate the formation of highly ordered activated complexes with a loss in entropy.

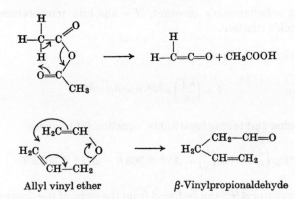

Allyl vinyl ether   β-Vinylpropionaldehyde

Simple bimolecular gas-phase reactions generally have small positive entropies of activation (see Table 6.2). Again in qualitative terms we can regard the coming together of two molecules to form a single molecule of activated complex as the formation of a more ordered state.

In the case where dimerization gives cyclic products as in the following Diels–Alder reactions:

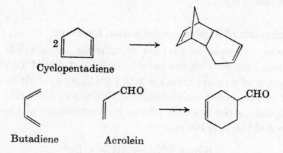

there is a large negative entropy of activation, qualitatively regarded as a loss in randomness in forming the cyclic intermediates.

Table 6.2. Bimolecular gas-phase reactions

	$\Delta S^{\ddagger}$ (J °K^{-1} mole^{-1})
HI + n-C$_3$H$_7$I	8.4
Cyclopentadiene dimerization	−108.7
Butadiene + acrolein	−83.6

When reactions are carried out in solution the effect of solvation produces quite large entropy effects. The orientation of solvent molecules around an ion is highly ordered, and the formation of ions from non-ionic species may be associated with a substantial decrease in entropy.

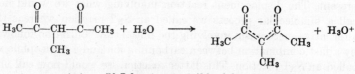

$\Delta S^{\ddagger} = -70$ J deg^{-1} mole^{-1} (16 cal deg^{-1} mole^{-1})

Such examples show how a knowledge of the entropy of activation can be of help in an understanding of reaction mechanisms. But caution must be used in the application of these principles, particularly for reactions in solution, because in such cases entropy effects also include changes in the arrangements of the solvent molecules as new species requiring differing degrees of solvation are formed from the reactants.

The Mechanism of Aliphatic Substitution Reactions

A classic example of the use of kinetics to establish reaction mechanisms occurs with the substitution reactions of alkyl halides. The reaction of an alkyl bromide with potassium cyanide in ethanolic solution shows second-order kinetics, the rate of the reaction being proportional to the product of the concentrations of the starting bromide and the cyanide ion.

$$RBr + CN^- \rightarrow RCN + Br^-$$

$$Rate = k[RBr][CN^-]$$

In many substitution reactions, however, the nucleophile is present in vast excess so that the consumption of the nucleophile during the reaction is negligible compared to its total concentration. The observed kinetics under these conditions will be first-order, and such a reaction is called a pseudo-first-order reaction.

$$RBr + \quad OH^- \quad \rightarrow ROH + Br^-$$
$$\text{(vast excess)}$$

$$Rate = k'[RBr] \quad \text{(pseudo-first-order)}$$

At the end of Chapter 3 of Part 1 we discussed reactions of alkyl halides which overall involved displacement but which were initiated by ionization of the carbon–halogen bond to form a transient carbonium ion, which subsequently reacted with some other nucleophile present. The displacement reaction involving what we would now call a bimolecular process we called an S_N2 reaction whereas the alternative mechanism involving the initial slow ionization followed by a fast combination between carbonium ion and a nucleophile we called an S_N1 reaction. This latter reaction we would now call unimolecular.

$$Y:^- + RX \rightarrow [Y\cdots R\cdots X]^- \rightarrow Y\!-\!R + X:^-$$
$$\text{Activated complex}$$

$$Rate = k[Y^-][RX] \quad \text{(If Y is in great excess, Rate} = k'[RX])$$

Bimolecular reaction, designated 'S_N2' or a 'displacement reaction'.

$$R—X \xrightleftharpoons{\text{slow}} R^+ + X:^-$$

$$R^+ + Y:^- \xrightarrow{\text{fast}} R—Y$$

Rate $= k[RX]$ (Rate completely independent of concentration of Y)

Unimolecular reaction, designated 'S_N1'.

The first kinetic evidence for the S_N1 mechanism is therefore that the reaction rate is independent of the concentration of nucleophile. Such kinetic behaviour is often observed for nucleophilic substitution when an intermediate carbonium ion can be stabilized by resonance or inductive effects. For instance, the rate of hydrolysis of t-butyl chloride in aqueous solution depends only on the concentration of t-butyl chloride. Added hydroxide, ethoxide, or acetate ion, although producing respectively the alcohol, ether, and acetate, has no effect on the rate, true first-order kinetics being observed:

$(CH_3)_3CCl + \bar{O}H \longrightarrow (CH_3)_3COH$
t-Butyl
chloride

$+ C_2H_5O^- \longrightarrow (CH_3)_3COC_2H_5$

$+ CH_3COO^- \longrightarrow (CH_3)_3COCOCH_3$

Rate $= k[(CH_3)_3CCl]$

Even more striking is the observation that diphenylmethyl chloride (benzhydryl chloride) reacts with fluoride ions, pyridine, and triethylamine at a rate that is not only independent of the concentration of the nucleophile but is also independent of the nature of the nucleophile.

$(C_6H_5)_2CHCl + F^- \xrightarrow{\text{liq.SO}_2} (C_6H_5)_2CHF + Cl^-$
Benzhydryl
chloride

$+ \underset{N}{\bigcirc} \xrightarrow{\text{liq.SO}_2} (C_6H_5)_2CH\overset{+}{N}\bigcirc Cl^-$

$+ (C_2H_5)_3N \xrightarrow{\text{liq.SO}_2} (C_6H_5)_2CH\overset{+}{N}(C_2H_5)_3\ Cl^-$

Rate $= k[(C_6H_5)_2CHCl]$

Such a result implies that the rate-determining step involves the benzhydryl chloride only.

However, it will be clear that the fact that a reaction shows first-order kinetics is no proof that it is truly unimolecular. The observed first-order behaviour may be entirely due to a pseudo-first-order reaction of the type we have described. Notice, however, that in all our reaction schemes for the S_N1 reaction we have drawn the first step as reversible. Any kinetic equations we derive from the reaction therefore should make allowance for this reversibility. We shall consider the case of an alkyl halide such as t-butyl chloride reacting in an aqueous medium.

$$RCl \underset{k_{-1}}{\overset{k_1}{\rightleftharpoons}} R^+ + Cl^-$$

$$R^+ + H_2O \xrightarrow{\text{fast}} ROH + H^+$$

$$Rate = \frac{d[ROH]}{dt} = k_2[R^+]$$

H_2O is the solvent and is present in excess; k_2 is the pseudo-first-order rate constant. We have no immediate means of determining the concentration of the carbonium ion, $[R^+]$, though we can express the rate of its formation in terms of the rate constants and the concentrations of the other species:

$$\frac{d[R^+]}{dt} = k_1[RCl] - k_{-1}[R^+][Cl^-] - k_2[R^+]$$

At first sight this additional equation appears to get us no further. We still do not know the concentration of the carbonium ion, so solution of the rate equation appears impossible. However, the carbonium ion is extremely reactive, combining either with chloride ions or water molecules as fast as it is formed. Its concentration therefore remains very small and, after an initial period, steady. If this is the case then we are justified in assuming that $d[R^+]/dt = 0$, although $[R^+]$ itself is finite. Therefore,

$$\frac{d[R^+]}{dt} = 0 = k_1[RCl] - k_{-1}[R^+][Cl^-] - k_2[R^+]$$

The assumption that the concentration of a reactive intermediate in a reaction sequence is constant is called the *steady-state approximation*.

We can now determine the concentration of the carbonium ion in terms of the rate constants and the other concentrations. Therefore,

$$[R^+] = \frac{k_1[RCl]}{k_{-1}[Cl^-] + k_2}$$

and hence we obtain an expression for the rate of formation of the alcohol:

$$\frac{d[ROH]}{dt} = \frac{k_1 k_2 [RCl]}{k_{-1}[Cl^-] + k_2}$$

Notice that according to this expression the rate is proportional to the concentration of t-butyl chloride but is decreased as $[Cl^-]$ is increased. The experimental observation of this rate equation for benzhydryl chloride and similar compounds provides the strongest direct evidence for the S_N1 mechanism. Since it is obvious from the equation that the addition of extraneous chloride ion will decrease the rate at which substitution occurs, this effect is sometimes known as the common-ion effect.

Consider then the hydrolysis of benzhydryl chloride in a water–acetone mixture. The rate of the hydrolysis is independent of the concentration of hydroxide ion. This shows that the mechanism does not involve an S_N2 displacement by hydroxide ion, but it does not rule out S_N2 displacement by water, namely:

$$(C_6H_5)_2CHCl + H_2O \xrightarrow{\text{slow}} (C_6H_5)_2CH\overset{+}{O}H_2 + Cl^- \xrightarrow[\text{OH}^-]{\text{fast}}$$
$$(C_6H_5)_2CHOH + H_2O$$

However, the common ion effect can be used to solve this problem, because while it is found that the addition of many salts increases the rate (they make the medium more polar, so favouring formation of a carbonium ion) the addition of lithium chloride markedly slows the hydrolysis.

The common-ion effect is therefore unambiguous evidence that this hydrolysis has an S_N1 mechanism. We must note, however, that this effect can only be observed if the carbonium ion intermediate is stable enough so that some of the carbonium ions are trapped by common ion rather than by solvent.

We have now as much evidence as kinetics alone can provide about the mechanisms of these reactions of alkyl halides. To obtain further information we still use kinetic methods but in conjunction with other observations.

We saw in Chapter 24 of Part 2 that the displacement reaction, as we had depicted, necessarily involves an inversion of configuration at the attacked carbon atom. The experimental confirmation of this aspect of the S_N2 mechanism was obtained by Hughes and his co-workers by showing that every displacement involved an inversion. They did this by taking an optically active alkyl iodide and treating it with potassium iodide in which the iodide ions were radioactive, i.e. 'labelled'. The rate of exchange was determined by separating organic and inorganic iodides and determining their radioactivity.

$$RI + I^{*-} \underset{k_{-e}}{\overset{k_e}{\rightleftharpoons}} RI^* + I^-$$

$$\frac{d[RI^*]}{dt} = k_e[RI][I^{*-}] - k_{-e}[RI^*][I^-]$$

where I^* represents radioactive iodine, and $k_e = k_{-e}$ is the rate constant for the exchange reaction.

It had previously been shown that the rate of racemization was proportional to the iodide ion concentration. Hughes and his co-workers measured the rate of racemization at the same time as the exchange reaction.

$$(+)\text{-}RI + I^- \underset{k_{-i}}{\overset{k_i}{\rightleftharpoons}} (-)\text{-}RI + I^-$$

where $(+)$-RI represents dextrorotatory alkyl iodide, and $k_i = k_{-i}$ is the rate constant for the inversion reaction.

$$\frac{-d[(+)\text{-}RI]}{dt} = \frac{+d[(-)\text{-}RI]}{dt}$$

$$= k_i[(+)\text{-}RI][I^-] - k_{-i}[(-)\text{-}RI][I^-]$$

α, the specific rotation $= m([(+)\text{-}RI] - [(-)\text{-}RI])$, where m is a constant. Therefore,

$$\frac{d\alpha}{dt} = -2k_i[I^-]\alpha$$

If every displacement is accompanied by inversion, the rate of exchange will be identical with the rate of inversion; if not, the former value will be greater. The data of Table 6.3 leave, therefore, little room for doubt that a halide–halide ion displacement is always accompanied by inversion. At present, this principle is thought to apply to all S_N2 reactions on carbon.

Table 6.3.

Alkyl halide	Temp. (°C)	$k_e \times 10^4$ (l.mole^{-1} sec^{-1})	$k_i \times 10^4$ (l.mole^{-1} sec^{-1})
C_6H_{13}—$\overset{\displaystyle CH_3}{\underset{\displaystyle H}{C}}$—I	30	13.8 ± 1.1	13.1 ± 0.1
C_6H_5—$\overset{\displaystyle C_2H_5}{\underset{\displaystyle H}{C}}$—Br	30	8.72 ± 0.92	7.95 ± 0.12
CH_3—$\overset{\displaystyle CO_2Et}{\underset{\displaystyle H}{C}}$—Br	22	5.15 ± 0.50	5.24 ± 0.05

In most cases both S_N1 and S_N2 mechanisms operate simultaneously. Therefore the observed rate law, say for the hydrolysis of an alkyl halide, is usually of the form:

$$\text{Rate} = k_1[\text{RCl}] + k_2[\text{RCl}][\text{OH}^-]$$

k_1 increases from primary to tertiary alkyl halides, while k_2 decreases. We can therefore draw a diagram (see Figure 6.3) representing the effect of different alkyl groups on rate and mechanism in a given kind of nucleophilic substitution in alkyl compounds, for example, the alkaline hydrolysis of alkyl bromides, the medium and the temperature being always the same. As we pass along the series (methyl, ethyl, isopropyl, t-butyl) there is a gradual change in the importance of the two mechanisms, and corresponding to this change there will be observable changes, first in the absolute reaction rate,

and secondly in the kinetic form of the reaction, as can be understood
with the aid of Figure 6.3. This introduces an important concept
with which we are well familiar, namely, that the introduction of
substituents into a molecule can alter the reaction mechanism.

Another factor which can also have a similar effect on such a
reaction is the nucleophilicity of the anion. Since a nucleophile attacks
by using a pair of its own electrons, we should expect the most
effective nucleophile to be the one whose attacking atom has the

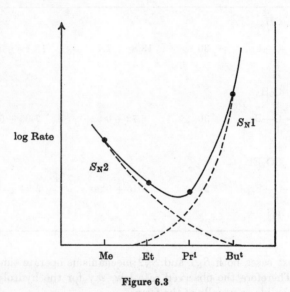

Figure 6.3

valence electrons most available for coordination. Broadly speaking,
this is the criterion for base strength, and it seems reasonable to
suggest that the strongest bases should be the most effective reagents for nucleophilic substitution reactions (see also Chapter 5).
However, basicity measures the tendency to bond to hydrogen, and
carbon 'basicity' or nucleophilicity need not parallel hydrogen
basicity. For example, it is common knowledge that such basic
species as phenoxide or acetate ions are less nucleophilic than a
non-basic ion such as I^-. In fact, several factors are involved in
determining nucleophilicity. First, when the attacking atom is the

same, e.g. oxygen or nitrogen, basicity does play a role; consequently we find the order, e.g.

ethoxide > phenoxide > acetate > nitrate.

Secondly, larger atoms are more nucleophilic than smaller ones, thus a thiophenoxide ion is a much better nucleophile than the phenoxide ion even though the latter is more basic. This size effect has been partly attributed to polarizability. A highly polarizable ion will have outer-shell electrons that are rather loosely held, and so they may more easily move in response to some demand. Thus the halide ions are in the order $I^- > Br^- > Cl^- > F^-$, the larger and more polarizable ones being the better nucleophiles. This is the same as saying that I^- is a soft base (Chapter 5, p. 231).

Larger atoms also tend to form hydrogen bonds poorly with solvent hydroxyls, so fewer bonds to the solvent need to be broken during the substitution reaction. This factor means that relative nucleophilicities in hydroxylic and non-hydroxylic solvents may be very different; for example, I^- is less nucleophilic than Br^- in acetone solution.

Therefore, if we examine a series of substitutions, by different nucleophilic substituting agents on the same alkyl derivative, under similar conditions of concentration, solvent, and temperature, and if we proceed through a series of nucleophilic reagents arranged in order of decreasing affinity for a carbon nucleus, we find that a powerful nucleophile such as hydroxide ion can initiate a bimolecular process whereas a weak nucleophile like a chloride ion will not attack the undissociated molecule but will react rapidly with a carbonium ion. We can expect therefore a region of mechanistic change, and observationally we can detect it, both by the change of kinetic form and by the change in the trend of the absolute rates. Such a change in mechanism is observed in the reaction of various anions with trimethylsulphonium ions in ethyl alcohol (see Table 6.4). Once again we can represent this result diagrammatically (see Figure 6.4).

We saw in Chapter 24 of Part 2 that although the S_N2 mechanism leads to complete inversion, the S_N1 reaction mechanism does not in fact yield complete racemization. We distinguished between those cases where a very stable carbonium ion was formed and almost complete racemization was observed and those cases in which an

Table 6.4. The effect of the nucleophile on reaction mechanism*

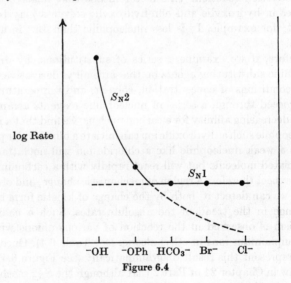

	S_N2	S_N1
	$k_2 \times 10^5$	$k_1 \times 10^5$
Anion	(l.mole^{-1} sec^{-1})	(sec^{-1})
OH$^-$	74,300	
C$_6$H$_5$O$^-$	1340	
HCO$_3^-$		7.38
Br$^-$		7.8
Cl$^-$		7.8

* It has been pointed out that the results in this table could also be interpreted in terms of a constant rate of solvent attack supervening in the reactions of the very weak nucleophiles.

Figure 6.4

unstable carbonium ion was involved and in which extensive inversion of configuration was observed. We argued that, in the cases involving the less stable carbonium ion, ionization produces an ion-pair which is still asymmetric even though the carbonium ion itself is planar. Thus, a nucleophile attacks more readily from the side of the carbonium ion away from the departing halide ion. The extent of inversion and racemization in these reactions, besides depending on the stability of the carbonium ion, also depends on the reactivity and solvating powers of the solvent. Ion pairs are most often encountered in acetic acid and similar solvents of high dielectric constant which promote ionization but are not very effective in solvating the free ions.

This ion-pair concept allows us to account for several recently observed phenomena which are difficult to reconcile with the simple S_N1 mechanism. The ion-pair of the type we have described is called an intimate ion-pair. When the carbonium ion is sufficiently stable the ion-pair may continue to exist even after one or two solvent molecules have come between the two ions. Such an entity is called a solvent-separated ion-pair, and it represents the stage preceding complete dissociation.

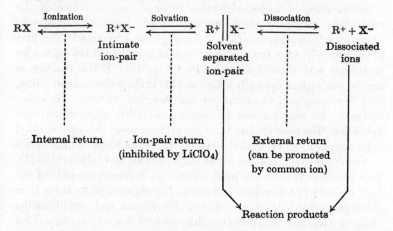

As indicated, the object of this hypothesis is to provide two distinguishable types of intermediate to explain the kinetic results. Thus the concept has been used to explain amongst other phenomena the so-called 'special salt effect' where the addition of a small concentration of a soluble salt such as lithium perchlorate brings about a marked increase in the rate of solvolysis but not in the rate of ionization. The perchlorate anion is believed to intervene in the ion-pair scheme, converting the ion-pair $R^+\|X^-$ into another ion-pair

written as R+||ClO₄⁻. This change reduces the likelihood of the reversal
to an intimate ion-pair, thus increasing the overall rate of solvolysis.

The most important feature of the S_N1 mechanism is that it fits
into our general picture of mechanistic organic chemistry. The
compounds which appear to undergo unimolecular solvolysis most
readily are the very compounds we should expect to yield the most
stable carbonium ions. The difficulty of the concept arises largely
from the use of the term 'unimolecular'. Clearly, a reaction of this
kind must involve a solvent, and in this sense the S_N1 reaction in
solution is quite different from the unimolecular reactions in the gas
phase we discussed originally. Difficulty has also arisen in attempts
to try to make the distinction between the S_N1 and the S_N2 reactions
completely rigid. Once you accept that the solvent takes part in the
S_N1 reaction, i.e. that it is not a truly unimolecular reaction at all,
then a lot of the conceptual difficulties disappear although the problems of explaining the observed kinetics remain.

Before going on to consider other types of reaction which can be
studied by kinetic methods we shall briefly consider two special
kinds of substitution reaction in an aliphatic compound. The first
of these involves the conversion of an alkanol into a chloroalkane by
treatment with thionyl chloride (Part 1, p. 184). If this reaction is
carried out with an optically active alcohol in the presence of pyridine,
both inversion and racemization are observed. However, in other
solvents, the reaction can be carried out with retention of configuration. The reaction has been shown to proceed through an alkyl
chlorosulphite, an ester that, with care, can be isolated and shown
to decompose upon heating to the alkyl chloride and sulphur dioxide.
It is assumed that in the inert solvent an intimate ion-pair of the
kind we have just discussed is formed. The chlorosulphite anion then
decomposes into sulphur dioxide and chloride ion, and recombination
of the new ion-pair then gives the chloride with the same configuration
as the original alcohol.

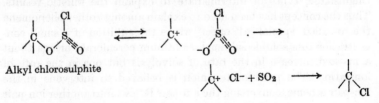

Alkyl chlorosulphite

In pyridine the ion-pair dissociates leading to partial inversion. This mechanism has been given the somewhat unsuitable symbol $S_N i$ (substitution, nucleophilic, internal).

Allyl halides undergo substitution reactions particularly readily by way of the stable allylic carbonium ion. Typically, the solvolysis of crotyl chloride (1-chlorobut-2-ene) in aqueous acetone yields a mixture of two isomeric alcohols:

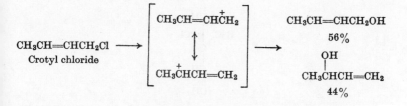

Such a reaction is sometimes known as an allylic rearrangement. Evidence that this kind of reaction can involve an intimate ion-pair of the type we discussed earlier comes from the solvolysis of γ,γ-dimethylallyl chloride (1-chloro-3-methylbut-2-ene) which, besides yielding the expected solvolysis products, yields at the same time a substantial proportion of the isomeric 3-chloro-3-methylbut-1-ene.

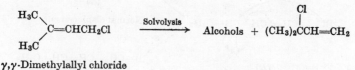

γ,γ-Dimethylallyl chloride

No common-ion effect is observed, and if the reaction is carried out in the presence of radioactive chloride, isomerization is found to proceed much more rapidly than the incorporation of labelled chloride. Again it is attractive to propose an ion-pair which collapses to yield an isomeric chloride instead of reacting with an external nucleophile. This mechanism is called an internal return and is given the unfortunate symbol $S_N i'$.

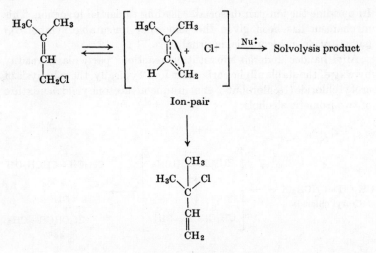

$S_N i'$ mechanism

It should not be thought that all substitution reactions of allylic compounds proceed by unimolecular processes. Normal S_N2 reactions occur with the expected inversion of configuration. However, conjugate substitution may also occur as depicted in the following sequence.

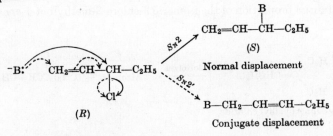

In the particular example illustrated, if the attacking anion B is the anion derived from diethyl malonate, 77% of the product is the result of normal S_N2 displacement and 23% of the product is the result of conjugate displacement (the so-called S_N2' reaction).

The Mechanism of Aliphatic Elimination Reactions

In Part 1 we followed a discussion of aliphatic substitution reactions (Chapter 3, p. 20) by a discussion of aliphatic elimination reactions

(Chapter 5, p. 43). The results of a kinetic study of olefin-forming elimination reactions are very similar to the kinetic results in the aliphatic substitution reactions. That is to say there are those elimination reactions whose rate depends on the concentration of both the nucleophile and the alkyl halide and those reactions in which the rate of elimination depends only on the concentration of the alkyl halides.

$$\begin{array}{ccc} R_2CH & & R_2C \\ | & + Nu:^- \rightarrow & || + HNu + X:^- \\ R_2CX & & R_2C \end{array}$$

$$\text{Rate} = k[R_2CHCXR_2][Nu:^-]$$

Observed kinetics of bimolecular elimination, $E2$, and unimolecular elimination of the conjugate base, $E1cB$.

$$\begin{array}{ccc} R_2CH & \xrightleftharpoons[\text{}]{\text{slow}} & R_2CH \\ | & & | & + X:^- \\ R_2CX & & R_2C^+ \end{array}$$

$$\begin{array}{ccc} R_2CH & & R_2C \\ | & + Nu: \xrightarrow{\text{fast}} & || + NuH^+ \\ R_2C^+ & & R_2C \end{array}$$

$$\text{Rate} = k[R_2CHCXR_2]$$

Observed kinetics of unimolecular elimination, $E1$.

At first sight it would appear that we have a situation almost identical with the substitution reaction with a unimolecular elimination involving initial ionization to form a carbonium ion which, instead of taking up a nucleophile as in the S_N1 substitution reaction, loses a proton to form the olefin. We can reasonably call this an $E1$ reaction, and the same factors which favour S_N1 reactions, i.e. formation of a stabilized carbonium ion in highly polar media, will operate here. As we have indicated above, the evidence for an $E1$ mechanism is first and foremost kinetic because, since the rate-determining step does not involve the base, the reaction rate will be independent of base concentration. But most $E1$ eliminations accompany the solvolysis of tertiary alkyl halides in solvents containing no other base. Under these conditions an $E2$ reaction in which the solvent acts as the base would not be distinguishable kinetically from a true $E1$ reaction. In cases such as this a more subtle mechanistic criterion is needed. Often the composition of the product mixture produces the answer, because if the reaction follows an $E1$ mechanism it involves formation of a carbonium ion, and the behaviour of this carbonium ion is independent of the nature of the leaving group. Thus, in the competition between S_N1 and $E1$ reactions, a mixture of olefin and ether (if the solvent is ethanol) is formed, and the composition of this mixture is the same whether the starting material is, say, t-butyl chloride or t-butyldimethyl-sulphonium ion.

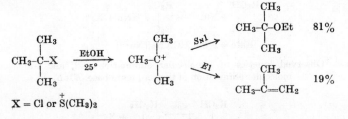

Another test which may be applied is the composition of the olefin mixture. Thus solvolysis of t-pentyl derivatives yields, in addition to substitution products, a mixture of two olefins, 2-methylbut-1-ene and 2-methylbut-2-ene. The more substituted olefin is preferred (this may be attributed to interaction of the C–H bonds of the alkyl substituent with the π electrons of the double bond; the more alkyl substituents there are, the more such interactions will be possible and the greater the stabilization resulting from them). Moreover, the observation is made that the proportion of the two olefins is independent of the leaving group, as illustrated for the case of t-pentyl bromide and dimethyl-t-pentylsulphonium ion, again indicating that a common intermediate is involved in the elimination.

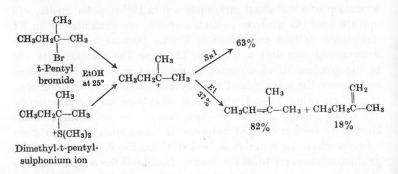

While an important diagnostic feature of eliminations by the $E1$ mechanism is that, aside from salt effects, their rates are not affected by the addition of bases, there are a considerable number of olefin-forming eliminations that are accelerated by base, many of these being exceedingly slow in the absence of base. For instance, when ethyl bromide is kept at 60° in ethanol, nothing happens because the ethyl cation is not stable enough for $E1$ elimination to occur. However, in the presence of a strong base such as sodium ethoxide, elimination of hydrogen bromide occurs and ethylene is formed (together with a considerable amount of diethyl ether from an S_N2 reaction). Since we find that the rate of olefin formation is proportional to both the ethyl bromide and the ethoxide ion concentrations, a bimolecular mechanism (the $E2$ mechanism) is written. This bimolecular mechanism we have frequently depicted in both Parts 1 and 2 as a concerted process.

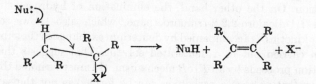

Bimolecular elimination depicted as a concerted process

Such a process would clearly exhibit second-order kinetics. However, there is another process in which the rate of reaction would depend on both the concentration of the nucleophile and the alkyl halide, and yet not be this concerted type. If the base were strong enough it is possible to conceive a system in which the reactant would be in equilibrium with its conjugate base.

$$\begin{array}{c} R_2CH \\ | \\ R_2CX \end{array} + B: \underset{\text{fast}}{\overset{\text{fast}}{\rightleftharpoons}} \begin{array}{c} R_2C:^- \\ | \\ R_2CX \end{array} + BH^+$$

Reactant   Conjugate base

The conjugate base could now lose a halide ion in a slow rate-determining process.

$$\begin{array}{c} R_2C:^- \\ | \\ R_2CX \end{array} \xrightarrow{\text{slow}} \begin{array}{c} R_2C \\ || \\ R_2C \end{array} + X:^-$$

This two-step process involving the initial formation of a carbanion has been designated an *E1cB* reaction (elimination, unimolecular, conjugate base). This mechanism involves a reversible first step, and if the reaction were conducted in the deuterated solvent (e.g. C_2H_5OD) then exchange between the solvent and the reactant should be observed so that after partial transformation into the olefin a proportion of the recovered reactant should be deuterated.

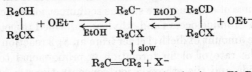

Concurrent deuterium exchange and elimination in an *E1cB* reaction

The reaction of phenethyl bromide with sodium ethoxide to produce styrene shows second-order kinetics but no deuterium exchange is observed, so this reaction is a true bimolecular elimination reaction. On the other hand, the elimination of hydrogen fluoride from 1,1-dichloro-2,2,2-trifluoroethane, which also shows second-order kinetics, is accompanied by deuterium exchange in the substrate when carried out in a deuterated solvent. This suggests that the reaction proceeds by an *E1cB* mechanism. (It must be noted that the demonstration that a carbanion can be formed is not the same as proof that it is an intermediate in the elimination.)

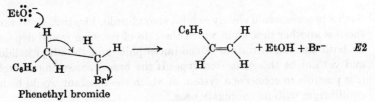

Phenethyl bromide

$$MeO^- + CF_3CHCl_2 \rightleftharpoons MeOH + CF_3\bar{C}Cl_2 \longrightarrow CF_2{=}CCl_2 + F^- \qquad \textit{E1cB}$$

1,1-Dichloro-2,2,2-
trifluoroethane

The *E2* mechanism is far more common than the *E1cB* reaction as far as alkyl halides are concerned, because if an *E1cB* reaction is to occur the leaving group has to depart from an intermediate carbanion in a slow step. But the principle of microscopic reversibility stipulates that, if a given sequence of steps constitutes the favoured mechanism for the forward reaction, the reverse sequence of these steps constitutes the favoured mechanism for the reverse reaction. It is apparent, however, that common leaving groups do not add to typical olefins, i.e. Y^- does not readily add to $>C=C<$. Consequently, $-\overset{|}{\underset{|}{C}}-\overset{|}{\underset{|}{C}}-Y$ is presumably unstable in most cases and is not formed from $H-\overset{|}{\underset{|}{C}}-\overset{|}{\underset{|}{C}}-Y$. Instead, concerted elimination of H and Y occurs.

Although the *E1cB* mechanism is restricted to olefins substituted with powerful electron-withdrawing groups, it can also be important in compounds of the type $HO-\overset{|}{\underset{|}{C}}-Y$, where Y is a powerful electron-withdrawing group. We are aware that, owing to the electronegativity of oxygen, addition of Y^- to $>C=O$ to give $\bar{O}-\overset{|}{\underset{|}{C}}-Y$ does occur. Moreover, $\bar{O}-\overset{|}{\underset{|}{C}}-Y$ is stabilized by the presence of Y so the *E1cB* mechanism can operate in the elimination of HY to form C=O bonds.

$$HO-\overset{|}{\underset{|}{C}}-Y \xrightarrow{\text{B:}} \bar{O}-\overset{|}{\underset{|}{C}}-Y \rightarrow O=\overset{|}{C} + Y^-$$

B: ≡ Base

(When Y is —CH_2CHO the reaction is a reverse aldol condensation.)

It is worth noting that formation of olefins by fragmentation reactions is also well established. Although the occurrence of these types of reaction is fairly general, γ-amino-halides have been used extensively as model compounds in mechanistic studies.

Two types of fragmentation have been recognised: (a) a two-step process where a carbonium ion is formed by a "normal" ionization process:

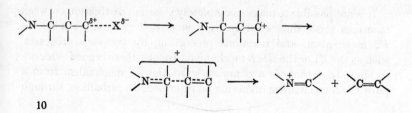

and (b) a one-step concerted process where ionization is assisted by electron release.

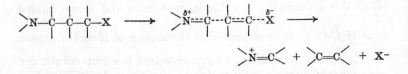

For concerted fragmentation as for *E*2 reactions, *trans*-antiparallel (antiperiplanar) geometry of the breaking bonds is the optimum geometry, as in:

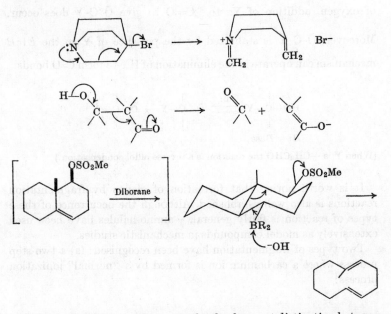

It is not possible to make a completely clear-cut distinction between reactions proceeding entirely by a concerted process, such as the *E*2 mechanism, and reactions proceeding by two-step processes, such as the *E*1 or the *E*1*cB* mechanism. In fact, there is good evidence to show that there is a complete gradation of mechanism from a pure *E*1*cB* mechanism involving an intermediate carbanion, through the true bimolecular process of the $E2$ mechanism, to the initial ionization forming a carbonium ion as depicted in the $E1$ mechanism. For reactions of the intermediate type we are concerned with the extent of bond-making and bond-breaking in the transition state.

$E1$-like $E2$-like $E1cB$-like

The kinetic distinctions between first- and second-order kinetics will not be sufficient to distinguish between such fine mechanistic differences but kinetic methods can be useful. The kinetic isotope effect provides us with a unique and often indispensable tool with which to investigate such reactions.

The Kinetic Isotope Effect

According to transition-state theory, introduced briefly in Chapter 8 of Part 2, reactants are in equilibrium with an 'activated complex' which subsequently decomposes in the rate-determining step to products. The activated complex differs from ordinary chemical molecules in lacking vibration along the line of the reaction path. Let us consider a reaction which involves the transfer of a hydrogen atom (H^+, $H\cdot$, or $H{:}^-$) from a carbon–hydrogen bond in a molecule R–H to another molecule B. To a first approximation, one of the degrees of freedom of R–H is the R–H stretching vibration; this vibration becomes a translation as the hydrogen atom moves from R to B.

The major portion of the kinetic isotope effect can be ascribed to changes in 'zero-point' energy when the reactants form the activated complex. We saw in Chapter 25 of Part 2 that the vibrational energy of a diatomic molecule was given by $E_v = h\nu(v + \frac{1}{2})$, where v (the vibrational quantum number) $= 0, 1, 2, 3, \ldots$. Thus, even at the lowest quantum level a diatomic molecule has a vibrational energy, the 'zero-point' energy ($E_0 = \frac{1}{2}h\nu$). We also saw in the same chapter that carbon–hydrogen bonds in large molecules could be regarded as bonds which have heavy masses (carbon residue) attached to light masses (hydrogen atom). The frequency of vibration of two masses joined by a spring is given by

$$\nu_H = \frac{1}{2\pi} \sqrt{\frac{k_H}{\mu_H}}$$

where k = force constant, μ = reduced mass.

Now,

$$\frac{1}{\mu_H} = \frac{1}{m_H} + \frac{1}{m_C}$$

But

$$m_C \gg m_H$$

Therefore,

$$\mu_H \approx 1$$

Similarly,

$$\mu_D \approx 2$$

To a good approximation,

$$k_H = k_D$$

Therefore,

$$\frac{\nu_H}{\nu_D} \approx \sqrt{2}$$

Thus, for the special case of the hydrogen atom attached to a large molecule being replaced by a deuterium atom, we have

$$\nu_H = \sqrt{2} \times \nu_D$$

Returning now to our chemical reaction, in the activated complex this stretching vibration becomes a degree of translational freedom and the vibration is lost. The difference in activation energy between a process involving a hydrogen atom and one involving a deuterium atom therefore corresponds approximately to the difference in the 'zero-point energy', i.e. $(\frac{1}{2}h\nu_{0H} - \frac{1}{2}h\nu_{0D})$, which is approximately 4.81 kJ mole^{-1} (1.15 kcal mole^{-1}). At 300°K this corresponds to a factor of about 7 in relative reaction rate. This large kinetic isotope effect is only observed for the isotopes of hydrogen (deuterium and tritium) where the mass ratios are greatest. Isotopic substitution of ^{13}C for ^{12}C may change a reaction rate by only a few per cent.

The kinetic isotope effect has been used to establish the $E2$ mechanism. In neither the $E1$ nor the $E1cB$ mechanism is the transfer of a hydrogen atom involved in the rate-determining step. In the true *E*2 mechanism, however, hydrogen (or deuterium) transfer is involved in the rate-determining process. The rate of elimination from a suitably deuterated 2-bromopropane in ethanolic sodium ethoxide has been studied, and an appreciably lower rate has been observed when a deuterium atom is being eliminated than when a proton is being removed (Table 6.5).

Table 6.5. Kinetic isotope effect in the *E*2 reaction

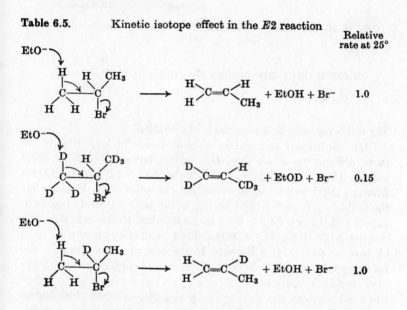

The isotope effect $k_H/k_D = 6.7$ is very close to the theoretical value for reactions at this temperature. For either an *E*1 mechanism or an *E*1c*B* mechanism this ratio should be unity. The simple theory of the kinetic isotope effect that we have described can account for the full effect, i.e. $k_H/k_D = 7$, or for no effect, i.e. $k_H/k_D = 1$. The interpretation of intermediate values for the kinetic isotope effects is in fact very difficult, and care should be taken not to assume that this necessarily means a partial breaking of the carbon–hydrogen bond in the activated complex.

Having illustrated the fact that apparently very similar *E*2 eliminations can involve the development of varying amounts of

carbanion and carbonium ion character, we may now be able to put
forward an explanation as to why $E2$ eliminations with neutral
substrates, such as alkyl halides and alkyl toluenesulphonates etc.,
ordinarily lead to the more substituted olefin (Saytzeff elimination)
whereas $E2$ elimination in a quaternary ammonium hydroxide yields
the least substituted olefin (Hofmann elimination).

Saytzeff   Hofmann

$$CH_3CHCH_2CH_3 + EtO^- \rightarrow CH_3CH{=}CHCH_3 + CH_3CH_2CH{=}CH_2$$
$$\quad\;\;|\qquad\qquad\qquad\qquad\quad 81\%\qquad\qquad 19\%$$
$$\quad\;\;Br$$

$$CH_3CHCH_2CH_3 + EtO^- \rightarrow CH_3CH{=}CHCH_3 + CH_3CH_2CH{=}CH_2$$
$$\quad\;\;|\qquad\qquad\qquad\qquad\quad 26\%\qquad\qquad 74\%$$
$$\quad\;\;{}^+S(CH_3)_2$$

The following may be a reasonable explanation.

Carbon–nitrogen and carbon–sulphur bonds in 'onium salts are
more difficult to break than the carbon–halogen bonds of alkyl
halides. Thus, as an alkoxide ion begins to remove a β-hydrogen atom
from an alkyl halide, some carbanion character may develop, but
the halide ion is such a good leaving group that it quickly begins to
depart and the new double bond starts forming. In this case then, the
reaction which forms the most stabilized olefin (alkyl groups stabilize
olefins; see page 274) will occur. In the case of an 'onium salt, the
leaving group is more difficult to displace, and consequently a rather
high electron density must be built up on the β-carbon atom before
the bond between the leaving group and the α-carbon atom begins
to break. The transition state, therefore, is one which has a considerable amount of carbanion character, and so, given the chance, the
reaction which involves the better carbanion will occur. (Primary
carbanions are more stable than secondary carbanions, and secondary
carbanions are more stable than tertiary carbanions, since alkyl
groups are electron-repelling with respect to hydrogen atoms.)

The observed variation of reaction path with the nature of the
leaving group is also consistent with a steric explanation. Consider,
for example, the elimination reaction of 1-methylbutyl derivatives,
where the formation of pent-1-ene would result from Hofmann-type
elimination and the formation of pent-2-ene would result from
Saytzeff-type elimination.

$$CH_3CH_2CH_2CHCH_3$$
$$\underset{X}{|}$$

$$\longrightarrow CH_3CH_2CH_2CH{=}CH_2 \quad \text{Hofmann type}$$
$$\longrightarrow CH_3CH_2CH{=}CHCH_3 \quad \text{Saytzeff type}$$

Experimental results show that the proportion of Hofmann-type product formed increases as the size of the leaving group (X) increases.

X =	Br	$\overset{+}{S}(CH_3)_2$	SO_2CH_3	$\overset{+}{N}(CH_3)_3$
$\dfrac{\text{Pent-1-ene}}{\text{Pent-2-ene}}$	0.45	6.7	7.7	50

Not only is the ratio dependent on the size of the leaving group but it is also dependent on the degree of branching of the hydrocarbon residue in the halide being treated.

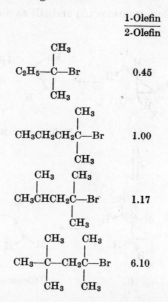

	$\dfrac{\text{1-Olefin}}{\text{2-Olefin}}$
$C_2H_5{-}\overset{\overset{\displaystyle CH_3}{|}}{\underset{\underset{\displaystyle CH_3}{|}}{C}}{-}Br$	0.45
$CH_3CH_2CH_2\overset{\overset{\displaystyle CH_3}{|}}{\underset{\underset{\displaystyle CH_3}{|}}{C}}{-}Br$	1.00
$CH_3\overset{\overset{\displaystyle CH_3}{|}}{C}HCH_2\overset{\overset{\displaystyle CH_3}{|}}{\underset{\underset{\displaystyle CH_3}{|}}{C}}{-}Br$	1.17
$CH_3{-}\overset{\overset{\displaystyle CH_3}{|}}{\underset{\underset{\displaystyle CH_3}{|}}{C}}{-}CH_2\overset{\overset{\displaystyle CH_3}{|}}{\underset{\underset{\displaystyle CH_3}{|}}{C}}{-}Br$	6.10

Thus, by increasing the degree of branching in the alkyl halide it is possible to disfavour the Saytzeff reaction path for these compounds.

The effect of increasing the bulk of the attacking base is often more striking. In the mixture of olefins resulting from the action of alkoxide bases on the bromide $(CH_3)_2CHC(CH_3)_2Br$, the ratio of 1-olefin to 2-olefin is shown.

Base	$C_2H_5O^-$	$(CH_3)_3CO^-$	$(CH_3)_2\overset{\overset{\displaystyle C_2H_5}{\displaystyle \vert}}{C}O^-$	$(C_2H_5)_3CO^-$
$\dfrac{\text{1-Olefin}}{\text{2-Olefin}}$	0.25	2.7	4.3	11.4

Examination of all these results shows that the shift from a Saytzeff-type to a Hofmann-type course is secured in each instance as the degree of steric crowding is increased, regardless of whether the crowding is in the leaving group, the alkyl bromide, or the attacking base. Consequently, a stereochemical picture to account for these facts must show the influence of each of these factors on the transition state for the elimination reaction. Consider the substrate

$$RCH_2C(CH_3)_2X$$

which has the two conformations (A) and (B) as shown.

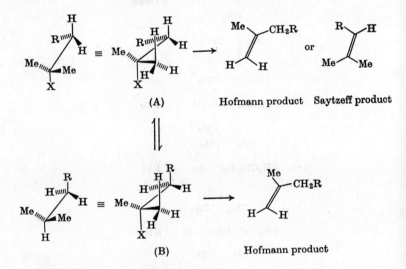

(A)   Hofmann product   Saytzeff product

(B)   Hofmann product

Assuming that antiperiplanar geometry is required, (B) can only give the Hofmann product since it has only one type of antiperiplanar hydrogen atom. (B) will be more favoured, i.e: the more preponderant conformer, the larger X and R are. In addition, Saytzeff elimination from (A) causes an increase in steric strain as R and Me become coplanar, while Hofmann elimination from (A) causes, probably, a reduction in strain. Consequently the bigger R is the greater the tendency for (A) to give Hofmann products. We can also see that attack by a bulky base on methyl in either (A) or (B) is hindered by Me and CH_2R, while attack by a bulky base on CH_2 in (A) is hindered by two Me groups and also by R. This latter attack is therefore less likely to occur.

Note that the toluene-*p*-sulphonyloxy group ($-OSO_2C_6H_4CH_3$-*p*) is bulkier than most of the other groups that we have discussed, yet, as a leaving group, its steric requirements are not great because it is not branched at the point of attachment, but one atom further away [cf. $\overset{+}{N}(CH_3)_3$]. Consequently, toluene-*p*-sulphonates undergo the same type of elimination reactions as do halides.

Catalysis

Often the manner in which a reaction can be accelerated, or inhibited, by certain added substances, provides a hint as to its mechanism. Thus, one of the most convincing ways of showing that a reaction proceeds through a path involving free radicals is the demonstration that the reaction is accelerated by substances such as peroxides and azo-compounds which readily produce free radicals, and inhibited by compounds such as hydrogen iodide and hydroquinone which are known to reduce the concentration of active free radicals.

Whereas an increase in temperature accelerates a reaction without normally altering its reaction path [at a higher temperature more molecules possess the energy required to reach the transition state (Part 2, p. 74)], a catalyst accelerates a reaction by supplying an alternative lower-energy path for reactants to change to products (see Figure 6.5), i.e. it changes the mechanism. For instance, the activation energy required for a molecule of methyl bromide to react with water to yield methyl alcohol, according to the single step

$$CH_3Br + H_2O \rightarrow CH_3OH + HBr$$

is so high that the reaction proceeds only very slowly. On the other hand, the activation energy for methyl bromide to react first with iodide ion (iodide is a very powerful nucleophile; see p. 267) and then for the methyl iodide thus formed to react with water (iodide is also a very effective leaving group because the very low C–I bond energy, 238 kJ mole⁻¹; cf. C–Br and C–OH bond energies of 272 and 330

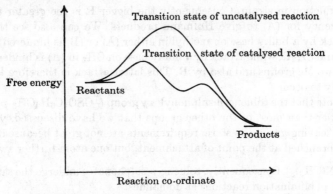

(N.B. Both curves may, of course, have several maxima)

Figure 6.5

kJ mole^{-1}, respectively, facilitates reaction) is low enough to permit the two-stage reaction

$$CH_3Br + I^- \rightarrow CH_3I + Br^-$$
$$CH_3I + H_2O \rightarrow CH_3OH + H^+ + I^-$$

to proceed much faster than the one-stage reaction. The dual ability of iodide to attack and depart readily in nucleophilic displacements therefore makes it an effective catalyst in such a reaction, the net result being the destruction of methyl bromide while the iodide is continually being used, then regenerated.

Acid and Base Catalysis

By far the most important cases of catalysis in organic chemistry involve the action of acids and bases. A common example occurs in the hydrolysis of carboxylic esters or amides. In Chapter 8 of Part 1 we depicted acid-catalysed esterification and hydrolysis. The function of the acid catalyst is to protonate either the acid or the ester to make it more susceptible to nucleophilic attack by alcohol or water. On page 88 in the same chapter we depicted base hydrolysis as involving nucleophilic addition of the hydroxide anion to the carbonyl bond. The kinetics of both acid and base hydrolysis of esters are consistent with these proposed mechanisms. Base hydrolysis shows second-order kinetics, and hydrolysis using water enriched with ^{18}O produces 'labelled' acid and 'unlabelled' alcohol as is required by the proposed mechanism.

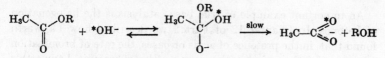

Base hydrolysis of an ester

*O represents ^{18}O (N.B. all the label appears in the acid, none in the alcohol)

Further evidence for the mechanism can be obtained by studying the hydrolysis of an ester in which the carbonyl oxygen atom is labelled. If such an ester is partially hydrolysed it is found that the ^{18}O content of the recovered ester has decreased. This is what we should expect from the mechanism shown below where a long-lived intermediate (A) is involved, because if in the exchange reaction the labelled oxygen must depart as *OH⁻, the lifetime of (A) must be sufficient to allow a significant chance for its isomerization to (B).

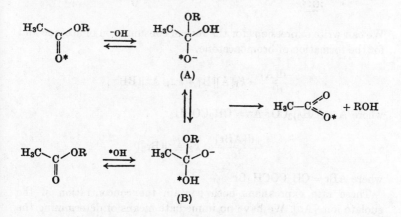

The result of such an experiment also rules out a direct, concerted displacement reaction.

So-called acid-catalysed hydrolysis of an ester is a true catalysis. The acid is not consumed in the reaction, and its sole function is to produce a positively charged intermediate which is more susceptible to attack by water as a nucleophile than is the unprotonated ester. Base hydrolysis of an ester is not really a catalysis at all because the base is consumed during the reaction.

An important example of true base catalysis is the halogenation of ketones. In Chapter 12 of Part 2 we described how Lapworth found that, in the presence of acids or bases, the rate of bromination of acetone is independent of the concentration of bromine. Lapworth's kinetic studies of this reaction mark the real beginning of physical organic chemistry and the understanding of organic reaction mechanisms. We will write out again the base-catalysed mechanism we proposed in Part 2.

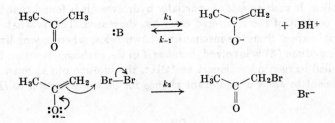

We can write expressions for the rate of consumption of acetone and for the formation of bromoacetone.

$$\frac{-d[A]}{dt} = k_1[A][B] - k_{-1}[A^-][BH^+]$$

where $A = (CH_3)_2CO$; $A^- = CH_3COCH_2^-$.

$$\frac{+d[ABr]}{dt} = k_2[A^-][Br_2]$$

where $ABr = CH_3COCH_2Br$.

These rate expressions both contain the concentration of the enolate ion, $[A^-]$. We have no immediate means of determining the concentration of the enolate ion though we can express the rate of its formation in terms of the rate constants and the concentration of the other species, namely:

$$\frac{d[A^-]}{dt} = k_1[A][B] - k_{-1}[A^-][BH^+] - k_2[A^-][Br_2]$$

At first sight this additional equation appears to get us no further. However, a solution may be derived by application of the steady-state approximation (see p. 262), i.e. it may be assumed that, very soon after the reaction gets under way, the concentration of the reactive intermediate A^- remains constant, i.e.

$$\frac{d[A^-]}{dt} = 0$$

We can now determine the concentration of the enolate anion in terms of the rate constants and the other concentrations. Therefore,

$$[A^-] = \frac{k_1[A][B]}{k_{-1}[BH^+] + k_2[Br_2]}$$

and hence we obtain an expression for the rate of formation of bromo-acetone:

$$\frac{d[ABr]}{dt} = \frac{k_1 k_2 [Br_2][A][B]}{k_{-1}[BH^+] + k_2[Br_2]}$$

Lapworth's observation was that the reaction is independent of bromine concentration, and this is compatible with the postulated mechanism provided that

$$k_2[Br_2] \gg k_{-1}[BH^+]$$

in which case,

$$\frac{d[ABr]}{dt} = k_1[A][B]$$

We have arrived at the rate equation for the reaction by assuming the intermediacy of an enolate ion, but it is most important to realize that, historically, Lapworth found that the reaction was independent of the bromine concentration. This observation then led him to introduce the hypothesis of an intermediate enolate anion.

We must now distinguish between *general* acid or base catalysis and *specific* acid or base catalysis. The most common kind of specific catalysis is that involving the conjugate acid or conjugate base of the solvent. In the case of water this would be the hydroxonium ion or the hydroxide ion. An example of specific hydroxonium ion catalysis is the hydrolysis of the acetal 1,1-diethoxyethane. The rate law for this reaction is

$$\frac{-d[Acetal]}{dt} = k[Acetal][H_3O^+]$$

It is found that the rate of this reaction is unaffected by the addition of other acids, say NH_4^+ ions, provided that the hydroxonium ion concentration (pH) remains constant. In other words, the acid catalysis is specific to the hydroxonium ion.

The base-catalysed bromination of some ketones, however, is found to be catalysed not only by hydroxide ions but also by chloroacetate ions even under conditions where both hydroxide ion concentration and total ionic strength are kept constant, showing that this reaction, in contrast to a specific base-catalysed reaction, is catalysed by chloroacetate ion (and by other bases) as well as by hydroxide ion. This reaction is therefore subject to general base catalysis. The criterion of distinction between a specific acid (or base) catalysis and a general acid (or base) catalysis is the behaviour of the rate when the concentrations of both components of a buffer solution are increased in the same proportion and at constant ionic strength. A buffer solution contains both H^+(aqueous) and an undissociated acid, and what the experiment sets out to do is to vary the concentration of the undissociated acid whilst keeping the concentration of H^+(aqueous) constant. This test for general or specific acid catalysis is shown in the (idealized) diagram.

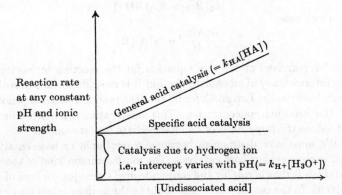

If the rate remains unchanged, the catalysis is of the specific acid (or base) type; if it increases, it is of the general acid (or base) type.

Since, in the bromination of ketones just discussed, the anion (^-OH or $ClCH_2CO_2^-$) is behaving as a base, we would not be surprised to discover that its catalytic effectiveness should be related to its base strength. This relationship is known as the Brønsted catalysis law.

$$\log k_c = \beta \log K_b + c \quad \text{(e.g. see Figure 6.6)}$$

where k_c is the catalysis constant for base B (i.e. it is the rate constant for a reaction catalysed by a base of ionization constant K_b, where $K_b = [BH^+][OH^-]/[B]$), β is the slope of the line, and c is the axial intercept. Therefore,

$$k_c = G(K_b)^\beta$$

where G is a constant characteristic of the reaction, solvent, and temperature. There is, of course, a similar law for general acid catalysis:

$$k_c = G(K_a)^\alpha; \qquad K_a = \frac{[A^-][H_3O^+]}{[HA]}$$

The first thing to notice about the Brønsted catalysis law is that there is nothing that we would not expect here. In fact, it would be rather surprising if there was no relationship between the dissociation constant of an acid and its effectiveness as a catalyst in an acid-catalysed reaction. The second thing to notice about the law is that we are relating a function of a rate constant (in this case the logarithm of a rate constant) to the same function of a dissociation constant; and the logarithm of a dissociation constant is directly proportional to the free energy change. Relationships of this kind are called 'linear free energy relations', about which we will have much more to say when we come to discuss the effect of substituents.

As an example of the use of the Brønsted relationship, let us consider the dehydration of acetaldehyde hydrate in acetone solution using a large number of carboxylic acids and phenols as catalysts. This reaction:

$$CH_3CH(OH)_2 \rightarrow CH_3CHO + H_2O$$
Acetaldehyde
hydrate

which can conveniently be followed dilatometrically by the increase in volume that occurs, is subject to general acid catalysis as well as general base catalysis. The mechanism for the acid catalysis is probably:

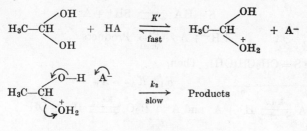

If k_2 is rate-determining, the above mechanism implies a relationship between k_2 and the basicity of A^-. Therefore, according to the Brønsted law,

$$\log k_2 = \beta \log K_{A^-} + G \tag{1}$$

However, if we plot the logarithms of the rate constant (k_A terms) for a particular acid HA against the $\log K_a$ for that acid, we get the following result (see Figure 6.6)

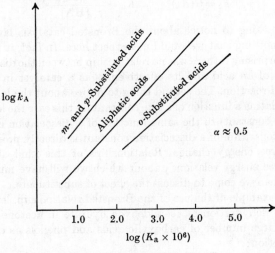

Figure 6.6

which implies the Brønsted relationship:

$$\log k_{obs} = \alpha \log K_{HA} + G' \tag{2}$$

However, expression (1) corresponds to expression (2) as shown below. Consider the mechanism:

$$S + HA \xrightleftharpoons{K'} SH^+ + A^-$$

$$SH^+ + A^- \xrightarrow[\text{(slow)}]{k_2} \text{Products}$$

where $S = CH_3CH(OH)_2$. Then,

$$K_{HA} K_{A^-} = K_w$$

for HA $\xrightleftharpoons{K_{HA}}$ $H^+ + A^-$ and $A^- + H_2O \xrightleftharpoons{K_{A^-}}$ HA + OH$^-$

Hence,

$$K_{A^-} = K_w / K_{HA}$$

According to the Brønsted law:

$$\log k_2 = \beta \log K_{A^-} + G$$
$$= \beta(\log K_w / K_{HA}) + G$$
$$= -14\beta - \beta \log K_{HA} + G \text{ (since } pK_w = 14)$$

Now,

$$K' = \frac{[SH^+][A^-]}{[S][HA]} = K_{HA} / K_{SH^+}$$

(K_{SH^+} from $SH^+ \rightleftharpoons S + H^+$)

I.e. $k_{obs} = k_2 K' = k_2 K_{HA} / K_{SH^+}$

Therefore,

$$\log k_{obs} = \log k_2 + \log K_{HA} - \log K_{SH^+}$$
$$= -14\beta - \beta \log K_{HA} + G + \log K_{HA} - \log K_{SH^+}$$
$$= (1 - \beta) \log K_{HA} + G'$$

since K_{SH^+} is constant for all acids HA.

I.e. $\log k_{obs} = \alpha \log K_{HA} + G'$

where $\alpha = (1 - \beta)$.

Since the observed value of $\alpha = {\sim}0.5$, $\beta = {\sim}0.5$ for the second step, and the observation of such a relationship demonstrates the applicability of the proposed mechanism.

One of the important features of the Brønsted equation is that, since the model process is the equilibrium between acids and their anions, the coefficient α must lie between 0 and 1, and is therefore a measure of the extent to which protonation or deprotonation has occurred in the transition state for the rate-limiting step. A value near unity represents a transition state resembling products; a value near zero indicates a resemblance to reactants.

Valuable information about the mechanism of acid-catalysed reactions is also provided by the use of acidity functions.

Acidity Functions

In many so-called acid-catalysed reactions the important process is the transfer of a proton from the reaction medium to the reacting molecule:

$$SH^+ + B \rightleftarrows BH^+ + S$$

where S is the solvent, SH^+ the solvated proton, and B the base or reactant. The equilibrium constant for this process (writing the equation the other way round) is given by:

$$K_{BH^+} = \frac{a_{SH^+}a_B}{a_{BH^+}}$$

where a_{SH^+}, a_B, and a_{BH^+} represent the activities of the species concerned. K_{BH^+} is the dissociation constant of the conjugate acid of B. Therefore,

$$pK_{BH^+} = -\log\frac{a_{SH^+}a_B}{a_{BH^+}}$$

In water,

$$pK_{BH^+} = -\log\frac{[H_3O^+][B]}{[BH^+]} - \log\frac{f_{H_3O^+}f_B}{f_{BH^+}}$$

where the f's are activity coefficients. In dilute aqueous solution,

$$pK_{BH^+} = -\log\frac{[H_3O^+][B]}{[BH^+]} \quad \left(\log\frac{f_{H_3O^+}f_B}{f_{BH^+}} = 0\right)$$

In order to determine K_{BH^+} we need to know $[H_3O^+]$, i.e. the pH, and also $[B]/[BH^+]$. This latter ratio can often be determined spectro-photometrically if there is a great difference in the visible absorption of B and its conjugate acid BH^+. If we now wish to know the dissociation constant of the conjugate acid of a weaker base C, we can do this by comparison:

$$pK_{CH^+} - pK_{BH^+} = -\log\frac{a_{SH^+}a_C}{a_{CH^+}} + \log\frac{a_{SH^+}a_B}{a_{BH^+}}$$

$$= \log\frac{[B]}{[BH^+]} - \log\frac{[C]}{[CH^+]} - \log\frac{f_C f_{BH^+}}{f_{CH^+}f_B}$$

Hammett has argued that in media of high dielectric constant the ratio multiplied by the inverse ratio of the activity coefficients of two similar mono-acidic bases will be close to unity, i.e.

$$\frac{f_C f_{BH^+}}{f_{CH^+}f_B} \approx 1$$

Therefore,

$$pK_{CH^+} - pK_{BH^+} = \log\frac{[B]}{[BH^+]} - \log\frac{[C]}{[CH^+]}$$

By employing a chain of such indicators Hammett was able to determine the pK_a's of some very weak bases, too weak to be studied in dilute aqueous media (Table 6.6).

Table 6.6. pK_a's of the conjugate acids of some very weak bases in strong sulphuric acid

4-Nitroaniline	+1.11
2,4-Dinitroaniline	−4.38
2,4,6-Trinitroaniline	−9.29

Let us now return to consider acid-catalysed reactions. If the reaction involves a pre-equilibrium in which the reactant molecule accepts a proton from the solvent, then it would be useful to be able to measure the proton-donating property of the medium. The acid dissociation constants are obviously not always of much value; for example, the pK_a of acetic acid is fairly low but a concentrated solution of this acid is obviously of very high acidity (i.e. proton-donating potential). Such a measure was devised by Hammett and is called an *acidity function*, H_0

$$H_0 = -\log\frac{a_{H^+}f_B}{f_{BH^+}}$$

Hammett suggested that, provided that B is a mono-acidic base and the medium has a high dielectric constant, the ratio f_B/f_{BH^+} would be the same for all bases, and if this is so H_0 becomes a property of the medium itself, i.e. the tendency of the medium to donate a proton to a neutral base. Notice that in dilute aqueous media H_0 becomes identical with pH. However, H_0 is meaningful in strong acid solutions where pH cannot be measured, and in entirely non-aqueous media such as acetic acid. H_0 for any solution can be determined by mono-acidic indicators whose pK_a's can be determined in the way we have already indicated.

$$H_0 = -\log\frac{a_{H^+}f_B}{f_{BH^+}}$$

$$= pK_a^{BH^+} + \log\frac{[B]}{[BH^+]}$$

These ideas which were developed in the 1930's have proved very fruitful, both in the study of acidic solutions and in the investigations of acid-catalysed reactions. However, it has become increasingly apparent that an H_0 scale depends on the nature of the indicators used in its determination. In other words, Hammett's assumption about the behaviour of activity coefficients has proved less accurate

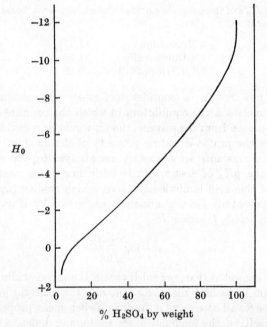

Figure 6.7. Acidity function, H_0, of H_2SO_4–H_2O solutions.

than was originally supposed. Figure 6.7 shows a plot of the sulphuric acid–water system ranging from water to 100% sulphuric acid. H_0 clearly exhibits the continuous increase in acidity with increasing acid concentration, the very great acidity of pure sulphuric acid being reflected by the large negative value of H_0 obtained (nearly -11). Determination of the H_0 values for mixtures of various acids with water (see Figure 6.8) shows that the two organic acids are appreciably weaker than the mineral acids. The crossover of the sulphuric acid curve above the perchloric acid curve at lower concentrations is probably due to contribution from the second dissociation constant.

The determination of the relative acidities of acids in very concentrated systems is of interest, but the real value of the acidity function is the use it can be put to in determining the mechanisms of reactions occurring in highly acidic media.

Remembering that H_0 measures the tendency of a solution to donate protons to an uncharged base, let us now consider a reaction

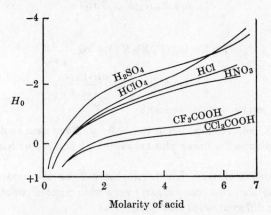

Molarity of acid

Figure 6.8. H_0 for aqueous solutions of acids.

which proceeds by a mechanism involving the initial fast protonation of a molecule.

$$X + H^+ \underset{\longleftarrow}{\overset{fast}{\longrightarrow}} HX^+$$

$$HX^+ \xrightarrow{slow} Products$$

Using the concepts of transition-state theory, we now have two equilibria, that between the substrate molecule and its conjugate acid, and that between the conjugate acid and the activated complex.

$$X + H^+ \underset{\longleftarrow}{\overset{K_H}{\longrightarrow}} XH^+ \underset{\longleftarrow}{\overset{K_\ddagger}{\longrightarrow}} (\ddagger)^\ddagger \xrightarrow{k_\ddagger} Products$$

The rate is given by

$$V = k_\ddagger[\ddagger]$$

$$K_\ddagger = \frac{[\ddagger]}{[HX^+]}\frac{f_\ddagger}{f_{HX^+}} \quad \text{and} \quad K_H = \frac{[XH^+]f_{XH^+}}{[X]f_X a_{H^+}}$$

Hence,

$$V = k_{\ddagger} K_H K_{\ddagger} \frac{f_X}{f_{\ddagger}} a_{H^+}[X]$$

The apparent overall first-order rate constant k is given by

$$V = -\frac{d[X]}{dt} = k[X]$$

Therefore,

$$k = k_{\ddagger} K_H K_{\ddagger} \frac{f_X}{f_{\ddagger}} a_{H^+}$$

Therefore,

$$\log k = \log(k_{\ddagger} K_H K_{\ddagger}) + \log \frac{a_{H^+} f_X}{f_{\ddagger}}$$

$$= \text{constant} + \log \frac{a_{H^+} f_X}{f_{\ddagger}}$$

i.e. $\log k = -H_0 + \text{constant}$

Thus, if X behaves like one of the bases we have used to define H_0, we should expect a linear plot between $\log k$ and H_0 with a slope of -1.

Just such a relation has been observed for the acid-catalysed hydrolysis of sucrose (cane sugar) over a wide range of concentrations using five different acids (see Figure 6.9.)

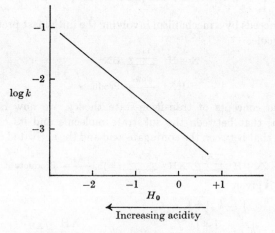

Figure 6.9

Let us now consider another type of acid-catalysed reaction, for example, the enolization of a ketone. We begin with a similar pre-equilibrium between the ketone and its conjugate acid but we then have a bimolecular process between the conjugate acid and a molecule of water to yield the enol.

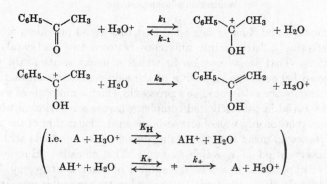

If we carry through exactly the same procedure as before we get the following expression for the rate:

$$V = k_{\pm} K_H K_{\pm} \frac{f_A}{f_{\pm}} a_{H_3O^+}[A]$$

so that as before the apparent first-order rate constant is given by:

$$k = k_{\pm} K_H K_{\pm} \frac{f_A}{f_{\pm}} a_{H_3O^+}$$

$$\log k = \log(\text{const.}) + \log[H_3O^+] + \log \frac{f_A f_{H_3O^+}}{f_{\pm}}$$

The last term in this equation, i.e. the log of the ratio of the activity coefficients, is likely to remain constant, so that in this case we should obtain a linear plot between the logarithm of the hydroxonium ion concentration and the logarithm of the rate constant. The difference between these two acid-catalysed reactions, the hydrolysis of sucrose, and the enolization of acetophenone is that in the enolization a molecule of water is involved in the activated complex.

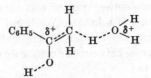

Activated complex in the acid-catalysed
enolization of acetophenone

Hammett and Zucker suggested that this would provide a means of detecting a fairly subtle difference between two acid-catalysed reactions. That is, a reaction in which a linear relationship was observed between the logarithm of the acid concentration and the logarithm of the rate would be a process in which a molecule of water was involved in the activated complex whereas a reaction in which a linear relationship was observed between the logarithm of the rate and H_0 would mean that the conjugate acid gave products without the intervention of a water molecule. This so-called 'Hammett–Zucker hypothesis' has been the subject of much controversy. The real difficulty is that not one but several water molecules seem to be involved in very many acid-catalysed reactions. In spite of the limitations of the Hammett–Zucker hypothesis, the use of acidity functions forms a vital part of the study of reactions occurring in strongly acidic media. In cases where a linear plot with a slope of -1 is obtained between $\log k$ and H_0, Hammett's arguments are still applicable.

Besides the H_0 scale, Hammett suggested H_+ and H_- scales, the former representing the donation of a proton to a positively charged species (not a very important function) and the latter representing the addition of a proton to a negatively charged species. This latter process is of course very common, and H_- scales have been utilized in mechanistic studies as have other acidity function scales.

Consider what happens when triphenylmethanol is dissolved in sulphuric acid. The triphenylmethyl cation is formed (Part 2, p. 108).

$$(C_6H_5)_3COH + 2H_2SO_4 \rightarrow (C_6H_5)_3\overset{+}{C} + 2HSO_4^- + H_3O^+$$

Triphenyl-
methanol

This can be represented in a general way as the protonation of a base to yield a cation and a molecule of water.

E.g. $$\text{ROH} + \text{H}^+ \rightleftharpoons \text{R}^+ + \text{H}_2\text{O}$$

$$K = \frac{a_{\text{R}^+}a_{\text{H}_2\text{O}}}{a_{\text{ROH}}a_{\text{H}^+}} = \frac{a_{\text{H}_2\text{O}}}{a_{\text{H}^+}}\frac{[\text{R}^+]}{[\text{ROH}]}\frac{f_{\text{R}^+}}{f_{\text{ROH}}}$$

This kind of base is quite common in organic chemistry, and it is therefore worthwhile introducing a new acidity function for such a system. This acidity function is given the symbol H_R.

$$H_\text{R} = \text{p}K_{\text{ROH}} + \log\frac{[\text{R}^+]}{[\text{ROH}]}$$

$$= -\log a_{\text{H}^+} + \log a_{\text{H}_2\text{O}} + \log\frac{f_{\text{R}^+}}{f_{\text{ROH}}}$$

Notice that there is a very simple relation between H_R and H_0:

$$H_\text{R} = H_0 + \log a_{\text{H}_2\text{O}}$$

so that these functions approach each other (and pH) as the solvent approaches pure water (Table 6.7).

Table 6.7. H_0 and H_R values of sulphuric acid solutions derived from ionization data for Ar_2CHOH and Ar_3COH

% H_2SO_4	H_0	H_R
5	−0.10	−0.07
20	−1.01	−1.92
60	−4.46	−8.92
80	−6.17	−14.12

Briefly then, correlations of the rate with H_0 and H_R are, respectively, diagnostic of ROH_2^+ and R^+ as the attacking species in a reaction, but this diagnosis can only be carried out in media in which H_0 and H_R differ significantly.

H_R has found its most important application in the study of the mechanism of aromatic nitration.

The Mechanism of Electrophilic Addition-with-Elimination Reactions, especially Nitration

In Chapter 14 of Part 1, we described how nitric acid, dissolved in concentrated sulphuric acid, behaves as a base. The nitroacidium ion tends to dissociate into the nitronium ion and a molecule of water. Although the equilibrium undoubtedly lies very much in favour of the nitroacidium ion, in the presence of concentrated sulphuric acid all the water is converted into the hydroxonium ion, so that the overall effect is one of complete conversion of nitric acid into the nitronium ion and the hydroxonium ion.

$$HNO_3 + H_2SO_4 \;\rightleftharpoons\; H_2NO_3^+ + HSO_4^-$$
$$\text{Nitroacidium}$$
$$\text{ion}$$

$$H_2NO_3^+ \;\rightleftharpoons\; NO_2^+ + H_2O$$
$$\text{Nitronium}$$
$$\text{ion}$$

$$H_2O + H_2SO_4 \;\rightleftharpoons\; H_3O^+ + HSO_4^-$$

$$\overline{HNO_3 + 2H_2SO_4 \;\rightleftharpoons\; NO_2^+ + H_3O^+ + 2HSO_4^-}$$

Evidence for this ionization comes principally from two sources. Pure sulphuric acid melts at 10.37°, and it is therefore possible to study the effect of added bases on the freezing point of sulphuric acid. Addition of water results in a van't Hoff depression factor of $i = 2$ due to the fact that each water molecule is protonated, so for every water molecule added to the sulphuric acid two species are formed, a hydroxonium ion and a hydrogen sulphate ion. Pure nitric acid added to sulphuric acid results in a van't Hoff depression of $i = 4$, one mole of nitric acid yielding four species, the nitronium ion, the hydroxonium ion, and two hydrogen sulphate ions. Additional evidence comes from spectroscopy. In Chapter 25 of Part 2 we briefly discussed Raman spectra and described how a Raman line is observed at 1400 cm^{-1} both in nitronium salts and in solutions of concentrated nitric acid in sulphuric acid. The ion responsible for this band has to be linear and triatomic, and the nitronium ion fits these requirements exactly.

Kinetic evidence that the nitronium ion is involved in the nitration of aromatic compounds comes from studies of the effect that acidity has on the rate of nitration. We have seen that nitric acid behaves as a base in sulphuric acid, and the conjugate acid decomposes to the nitronium ion and a molecule of water. In other words, nitric acid behaves in the same way as triphenylmethanol or other similar 'H_R bases'. Thus, the acidity function H_R should provide a measure of the concentration of the nitronium ion, and a plot of the logarithm of the rate constant against H_R should provide a straight line with a slope of unity.

The rate of nitration of compound ArH is given by

$$\text{Rate} = k[NO_2^+][ArH]$$

$$= k_2[HNO_3][ArH]$$

where k_2 varies with medium composition.

$$H_R = pK_{HNO_3} + \log \frac{[NO_2^+]}{[HNO_3]}$$

Therefore,

$$\log k_2 = (\log k - pK_{HNO_3}) + H_R$$

Lines of unit slope are observed when, for example, the rates of nitration of the phenyltrimethylammonium ion and the p-toluyltrimethylammonium ion are plotted against H_R.

In these reactions nitric acid cannot be the effective nitrating species because the variation of the rate of nitration when water is added is not related to the change in concentration of nitric acid, while if the nitroacidium ion were the reactive species the rate of nitration would follow the Hammett acidity function H_0 and not the H_R function, as shown above.

Having established that the rate of nitration is proportional to the concentration of the nitronium ion, we have proceeded as far as direct kinetic observations will take us. In Chapter 13 of Part 1, and again in Chapter 14, we depicted aromatic electrophilic substitution reactions as proceeding by initial addition of a cation to the aromatic nucleus. Ionic compounds, whose structure corresponds to that we have represented as the Wheland intermediate, have been isolated and are stable at low temperatures. Since in these compounds the acceptor group is linked by a σ bond to a particular aromatic carbon atom, they are called σ complexes. For example, when benzene is added to a mixture of hydrogen fluoride and boron trifluoride a reaction occurs with the formation of a highly coloured σ complex. A proton has been added to one of the benzene-ring carbons, and in the resulting species the proton and the ring are complexed by the formation of a new σ bond. The structures of such complexes have been supported by spectroscopic studies together

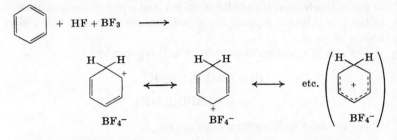

with the fact that they are electrically conducting, as would be expected of salts. Moreover, when deuterium fluoride (DF) is used instead of hydrogen fluoride, the deuterium exchanges with the benzene protons. This is consistent with the structure assigned to the σ complex because the deuterium newly attached to the ring carbon is equivalent to the proton already on that benzene carbon, and consequently either may be lost when the complex is decomposed.

Regarding nitration, we find that nitronium fluoroborate ($NO_2^+BF_4^-$) is an effective nitrating agent, and toluene treated with nitronium fluoroborate at low temperatures in a suitable solvent yields the crystalline ionic σ complex shown.

The isolation of such an ionic compound is not in itself evidence that such an intermediate actually occurs in ordinary nitration. However, taken in conjunction with the kinetic evidence it provides strong evidence that something very akin to the isolated ionic intermediate is involved.

We can now depict nitration by the nitronium ion and similar electrophilic addition-with-elimination reactions as follows:

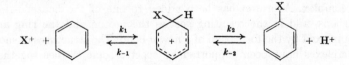

Nitration is not a reversible reaction, so either k_{-1} or much more likely k_{-2} must be very small indeed. The rate of nitration of nitrobenzene and some other similar compounds actually decreases in the very strongest sulphuric acid media. It was at one time suggested that this was because k_2 was rate-determining and that in very strong acid the concentration of base to accept the proton decreased. However, a study of the rate of nitration of fully deuterated compounds showed that there was no kinetic isotope effect, so clearly, in the nitration, k_1 is rate-determining and the transfer of a proton is fast. It would be quite wrong to conclude that this was true for all aromatic electrophilic addition-with-elimination reactions. Two important reactions show a kinetic isotope effect. These are the coupling of aromatic diazonium salts with certain phenoxide anions and aromatic sulphonation. In diazonium coupling we are dealing with a very weak electrophile and k_{-1} is probably of considerable magnitude, that is to say, the first equilibrium may lie on the side of the reactant, in which case the formation of the coupled product depends on k_2. The actual nature of the electrophile in aromatic sulphonation is still a matter of some controversy, though there is little doubt that sulphur trioxide is involved, probably associated with the solvent in some way. Aromatic sulphonation is completely reversible, so again it is easy to see why k_2 can have a rate-determining effect.

We must return briefly to nitration in order to consider the nitration of more reactive aromatic compounds in less acidic media. We have indicated that a nitronium ion only occurs in highly acidic media, and yet phenol and dimethylaniline can be nitrated in dilute aqueous media. Further, there is a large range of intermediate compounds such as polyalkylbenzenes and phenyl ethers such as anisole which can be nitrated by solutions of nitric acid in acetic acid or in nitromethane. The nitration of mesitylene in nitromethane in which nitric acid is in a large excess shows pseudo-zero-order kinetics, i.e. the reaction is independent of the concentration of the hydrocarbon. It was originally suggested that the rate-determining step was the self-ionization of nitric acid to yield the nitronium ion which then reacted rapidly. This interpretation has since been questioned and the exact nature of the nitration process in these organic solvents is at present uncertain. In dilute aqueous media the nitronium ion cannot exist, and nitration must therefore involve other species.

Nitration of phenols and amines is catalysed by nitrous acid normally present in trace amounts in nitric acid. Nitrous acid is a much stronger base than nitric acid but like nitric acid the nitrous acidium ion breaks down to give the nitrosonium ion and water, this process occurring at very much lower acidities than the formation of the nitronium ion.

$$H^+ + HNO_2 \rightleftharpoons H_2NO_2^+$$
Nitrous acidium ion

$$H_2NO_2^+ \rightleftharpoons NO^+ + H_2O$$
Nitrosonium ion

Nitrous acid catalysed nitration probably involves initial nitrosation by the nitrosonium ion followed by oxidation of the aromatic nitroso-compound so formed, to yield the nitro-compound and more nitrous acid. It is now known that nitric acid is not the oxidizing agent, and it is thought probable that it is N_2O_4 that is effective.

$$ArNO + N_2O_4 \rightarrow ArNO_2 + N_2O_3$$

In the intermediate range between dilute aqueous nitric acid and solutions of nitric acid and excess sulphuric acid, a variety of nitrating agents are possible. A particularly important one is likely to be the anhydride of nitric acid, N_2O_5. This molecule will be insufficiently electrophilic to attack nitrobenzene, but it may well be the nitrating agent for more reactive aromatic compounds.

$$2 HNO_3 \rightleftharpoons N_2O_5 + H_2O$$

$$O_2N\text{-}O\text{-}NO_2 + ArH \rightleftharpoons Ar\underset{H}{\overset{+}{\diagdown}}NO_2 + NO_3^- \longrightarrow ArNO_2 + HNO_3$$

An important feature to notice about aromatic nitration is that no single simple mechanism is sufficient to account for the nitration of the wide range of aromatic compounds under vastly different conditions. This plurality of reaction mechanisms is very common and could almost be said to be one of the basic concepts of mechanistic organic chemistry. Another simple case where a single mechanism is inadequate to account for the observed experimental results is the Friedel–Crafts acylation of aromatic nuclei. Reactive molecules such as anisole and mesitylene can be acylated to yield the corresponding aromatic methyl ketone by being treated with solutions of acetic acid in trifluoroacetic anhydride or by solutions of acetyl perchlorate. Both of these media can be shown to contain the acetylium cation, and by analogy with nitration it would seem reasonable to suppose that the acetylium ion is the attacking species.

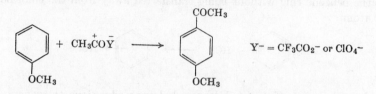

$Y^- = CF_3CO_2^-$ or ClO_4^-

Benzene, however, is unaffected by either of these systems, and yet benzene can be acylated by a mixture of acetyl chloride and aluminium chloride.

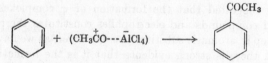

This reaction probably involves the bimolecular displacement of the aluminium tetrachloride anion from the complex. Aromatic compounds of intermediate reactivity such as toluene and the xylenes react slowly with solutions known to contain the acetylium ion and yet rapidly with the acetyl chloride–aluminium chloride complex. Probably in these cases both mechanisms are occurring simultaneously. This again is another characteristic of organic reactions. We have pointed out previously that in the reactions of alkyl halides with nucleophiles, displacement by the S_N1 and the S_N2 mechanisms may be occurring simultaneously, together with an elimination reaction, again possibly involving both $E1$ and $E2$ mechanisms occurring concurrently. In order to do the kind of mechanistic studies we have been describing it is necessary to find conditions in which one mechanism can be isolated from the others.

As well as forming σ complexes with molecules and ions capable of acting as electron-acceptors, benzenoid compounds also form π complexes in which the aromatic sextet as a whole acts as the electron donor and the electrophile forms a loose addition complex with the whole aromatic system, without being bonded to a particular carbon.

Thus when hydrogen chloride is dissolved in benzene it forms a
1:1 complex with the solvent. But the solution does not conduct
electricity so a salt has not been formed, nor does deuterium exchange
with the ring protons occur when DCl is used. Apparently the proton
of hydrogen chloride has a weak interaction with the π electrons of
the benzene ring without being transferred away from the chlorine
atom.

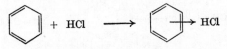

A great variety of electrophiles, e.g. halogens, silver ions, tetracyano-
ethylene, trinitrobenzenes, etc., give π complexes (often called
charge-transfer complexes) with aromatic substrates, and since
many electrophiles also carry out substitution on benzene rings it
has been suggested that the formation of π complexes between
benzenoid compounds and electrophiles constitutes the rate-determining step in aromatic substitutions. However, as we have already
indicated, there is strong evidence that it is the formation of the
σ complex and not the π complex which is the rate-determining step.

If we examine Table 6.8 we see that the halogenation rates of
various alkylbenzenes correlate very well with the relative stabilities

Table 6.8. Comparison of relative stabilities of complexes of alkylbenzenes
with bromination and nitration rates.

	Relative bromination rate	Relative nitration* rate	Relative σ complex stability (HF–BF$_3$)	Relative π complex stability (HCl)
Benzene	1	1	1	1
Toluene	605	1.67	7	1.51
p-Xylene	2500	1.96	11	1.65
o-Xylene	5300	1.75	12	1.85
m-Xylene	514,000	1.65	290	2.06
1,3,5-Trimethyl-benzene	189,000,000	2.71	145,000	2.60

* Using nitronium fluoroborate in tetramethylene sulphone.

of their σ complexes, but less well with the relative stabilities of their π complexes. We might expect this difference between the stability of σ and π complexes because in a π complex only a little charge is transferred to the ring since the principal resonance form is the neutral one. Also, the charge is placed in π orbitals delocalized over all six benzene carbon atoms so it can be stabilized more or less equally by methyl groups in any position. On the other hand, we would expect the relative stabilities of σ complexes to be strongly affected by both the number and the position of methyl substituents; thus in *m*-xylene the two *meta*-methyl groups, which do not have much effect on π complex stability, have a large effect on the stability of the σ complex, because the full electron-repelling effect of methyl groups *ortho* and *para* to the point of substitution is brought into play in the carbonium ion structure of the σ complex.

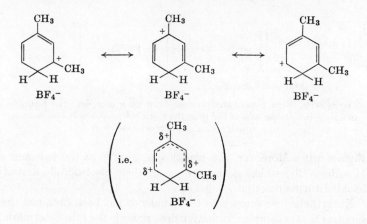

This greater response to the effect of substituents in electron-demanding reactions finds a parallel in many substitution reactions where *m*-xylene is invariably substituted faster than its isomers.

In addition to data such as those just outlined, and the previously mentioned isolation of σ complexes, practically all of which can, on heating, be converted into typical substitution products in high yields (see top of p. 310), the involvement of σ complexes rather than π complexes in the rate-determining step of aromatic electrophilic substitution reactions gains support from other studies. For instance, it is felt that the crystal structure of typical π complexes, as revealed

11

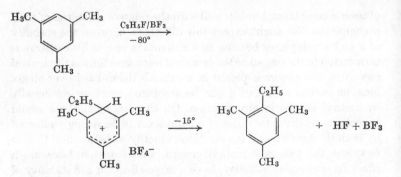

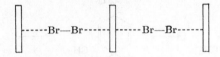

by X-ray diffraction studies, does not constitute a good model for aromatic substitution. For example, in the benzene–bromine complex the bromine lies above the centre of the benzene ring (see

Aromatic substrate

Figure 6.10. Structure of the benzene–bromine π complex; the bromine molecule lies on the axis of the ring, the nearest bromine atom interacting equally with all six p orbitals.

Figure 6.10). Moreover, the complexing ability of the halogens is $I_2 > Br_2 > Cl_2$, which is the reverse of their electrophilic activity in substitution reactions.

Nevertheless we must note that evidence has been obtained that suggests that π complex formation does provide the rate-determining step in certain substitutions. Thus we see from Table 6.8 that in certain nitration reactions the very small effects of the methyl groups are very similar to their effects on the stability of π complexes. In addition spectral evidence has been obtained for π complexes in some Friedel–Crafts reactions; for example, absorptions at 260 and 361 nm have been attributed to a 1:1:1 complex of toluene, t-butyl chloride, and boron trifluoride which is obtained when these three compounds are mixed at $-95°$.

Kinetic methods of examination can give information only about the maximum point on a free-energy profile of the reaction, and it is possible, and indeed likely, that π complexes are formed even in reactions in which the formation of a σ complex is rate-determining, and it may be the case that π complex formation provides the rate-determining step in certain substitutions. At present, the most general representation of the kinetic pathway of an aromatic substitution is one in which σ complex formation is rate-determining, and it is shown in Figure 6.11.

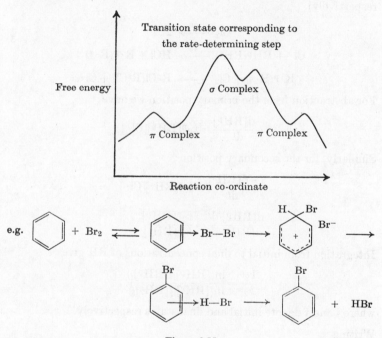

Figure 6.11

The Study of Free-radical Reactions

We saw in Chapter 11 of Part 2 that in the majority of free-radical reactions the concentration of radicals is extremely small but the reaction proceeds to completion because a chain mechanism is involved. Kinetic studies of these reactions nearly always depend on the use of the steady-state approximation. Let us begin with a very simple example. To measure the absolute rate of a radical reaction proceeding by a chain mechanism is extremely difficult, but to measure the relative rates of two competing radical reactions is fairly simple. Suppose for example we chlorinate n-butane in the gas phase and measure the amounts of 1- and 2-chlorobutane formed in the products at the end of the reaction. If we only carry the reaction out to a 1 or 2 per cent conversion, we can derive a very simple expression for the ratio of the rate constants for hydrogen abstraction at the primary and secondary positions (k^p and k^s respectively).

$$Cl_2 \xrightarrow{h\nu} 2Cl\cdot$$

$$Cl\cdot + RH^p(RH^s) \xrightarrow{k_2} HCl + R\cdot^p(R\cdot^s)$$

$$R\cdot^p(R\cdot^s) + Cl_2 \xrightarrow{k_3} R^pCl(R^sCl) + Cl\cdot$$

For abstraction from the primary position we have:

$$\frac{-d[RH^p]}{dt} = k_2^p[RH^p][Cl\cdot]$$

Similarly, for the secondary position:

$$\frac{-d[RH^s]}{dt} = k_2^s[RH^s][Cl\cdot]$$

$$\frac{d[RH^p]/dt}{d[RH^s]/dt} = \frac{k_2^p[RH^p]}{k_2^s[RH^s]}$$

Integration from initial to final concentrations of RH gives

$$\frac{k_2^p}{k_2^s} = \frac{\ln([RH^p]_i/[RH^p]_f)}{\ln([RH^s]_i/[RH^s]_f)}$$

where i and f denote initial and final states respectively.

Writing

$$\Delta RH^p = [RH^p]_i - [RH^p]_f$$

and

$$\Delta RH^s = [RH^s]_i - [RH^s]_f$$

we have

$$\frac{k_2^p}{k_2^s} = \frac{\ln(1 + \Delta RH^p/[RH^p]_f)}{\ln(1 + \Delta RH^s/[RH^s]_f)}$$

If the extent of reaction is small, then

$$\ln(1 + \Delta RH^p/[RH^p]_f) \approx \frac{\Delta RH^p}{[RH^p]_i}$$

Hence, for small conversion:

$$\frac{k_2^p}{k_2^s} = \frac{\Delta RH^p [RH^s]_i}{\Delta RH^s [RH^p]_i}$$

Now, ΔRH^p must equal the concentration of 1-chlorobutane (provided that there are no side-reactions and that termination steps are an unimportant way of removing RH), i.e. $\Delta RH^p = [RCl^p]_t$. Similarly,

$$\Delta RH^s = [RCl^s]_t$$

Hence

$$\frac{k_2^p}{k_2^s} = \frac{[RCl^p]_t}{[RCl^s]_t} \times \frac{[RH^s]_i}{[RH^p]_i}$$

When different sites in the same molecule are being studied the ratio $[RH^s]_i/[RH^p]_i$ is simply the ratio of the number of hydrogen atoms of each type in the molecule. The rate of attack per hydrogen atom in a single molecule is known as the 'relative selectivity' usually written RS.

For butane $RS_p^s = k_2^s/k_2^p \times 3/2$

The same equation can also be derived by a 'steady state' argument. This argument will apply not only in competition between two sites in the same molecule but also in competition between sites in different molecules, so that if we competitively chlorinate methane and propane together in the same reaction vessel we may treat the results in the same way. If therefore we know the absolute rate of a radical reaction it is often possible to measure other reactions going by a similar mechanism, competitively, and thereby obtain the absolute rate of a whole series of reactions. In the case of chlorination of alkanes the absolute rate has been determined for the reaction of chlorine with hydrogen, and all the subsequent data depend on competitive results, firstly on methane with hydrogen and subsequently methane with propane and so on.

Similar studies have been carried out using other atoms and radicals (see Table 6.9).

Another way of studying radical reactions is illustrated by the addition of bromotrichloromethane to ethylene. In the dark, no reaction occurs between these two molecules. Ethylene is unaffected by light but bromotrichloromethane is photolysed into trichloromethyl

Table 6.9. Hydrogen abstraction from halogenomethanes by atoms and radicals

$$RH + X\cdot \xrightarrow{k_2} R\cdot + HX$$

$k_2 = A_2 e^{-E_2/RT}$; A_2 in l. mole^{-1} sec^{-1}; E_2 in kJ mole^{-1}

RH	X = Cl		X = Br		X = CF$_3$		X = CH$_3$	
	$A_2 \times 10^{-10}$	E_2	$A_2 \times 10^{-10}$	E_2	$A_2 \times 10^{-7}$	E_2	$A_2 \times 10^{-7}$	E_2
H$_2$	4.1	23.1	14	82.7	1.7	37.0	400	51.2
CH$_4$	0.6	16.0	1.5	76.4	4	47.5	3	53.8
								(61.7)
CH$_3$Cl	1.1	13.9	1.3	60.1	6	44.5	60	39.5
CH$_2$Cl$_2$	1.3	12.6			0.6	31.9	20	30.2
CHCl$_3$	0.7	13.9	0.2	39.1	0.5	27.7	6	24.4
CH$_3$Br			1.7	66.8	4	43.7	320	42.4

radicals and bromine atoms. The subsequent chain reaction is as follows:

$$CCl_3Br \xrightarrow{h\nu} CCl_3\cdot + Br\cdot$$

$$CCl_3\cdot + CH_2{=}CH_2 \xrightarrow{k_2} CCl_3CH_2CH_2\cdot$$

$$CCl_3CH_2CH_2\cdot + CCl_3Br \xrightarrow{k_3} CCl_3CH_2CH_2Br + CCl_3\cdot$$

$$Br\cdot + CH_2{=}CH_2 \xrightarrow{k_4} BrCH_2CH_2\cdot$$

$$BrCH_2CH_2\cdot + CCl_3Br \xrightarrow{k_5} BrCH_2CH_2Br + CCl_3\cdot$$

$$CCl_3\cdot + CCl_3\cdot \xrightarrow{k_6} C_2Cl_6$$

Provided that bromotrichloromethane is in reasonable excess and the reaction is carried out to only a very small conversion, reaction (6), the combination of two trichloromethyl radicals, is the only appreciable chain-terminating step.

The steady-state approximation applied to the mechanism gives four conditions

$$\frac{d[CCl_3\cdot]}{dt} = 0, \quad \frac{d[Br\cdot]}{dt} = 0, \quad \frac{d[BrCH_2CH_2\cdot]}{dt} = 0,$$

$$\frac{d[CCl_3CH_2CH_2\cdot]}{dt} = 0$$

Expressing the rate of initiation by:

$$\text{Rate (i)} = \Phi I_a$$

where Φ is the quantum yield of bromotrichloromethane photolysis and I_a is the intensity of absorbed light, we have:

$$[CCl_3\cdot]^2 = \Phi I_a / k_6$$

$$[CCl_3CH_2CH_2\cdot] = \frac{k_2[CH_2{=}CH_2]}{k_3[CCl_3Br]}\left(\frac{\Phi I_a}{k_6}\right)^{1/2}$$

$$[Br\cdot] = \frac{\Phi I_a}{k_4[CH_2{=}CH_2]}$$

$$[BrCH_2CH_2\cdot] = \frac{k_4[CH_2{=}CH_2]\Phi I_a}{k_5[CCl_3Br]k_4[CH_2{=}CH_2]}$$

Then,

$$\frac{d[CCl_3CH_2CH_2Br]}{dt} = k_3[CCl_3CH_2CH_2\cdot][CCl_3Br]$$

$$= k_2[CH_2{=}CH_2](\Phi I_a / k_6)^{1/2}$$

$$\frac{d[C_2Cl_6]}{dt} = k_6[CCl_3\cdot]^2$$

$$= \Phi I_a$$

Experimentally, all that we can measure is the ratio of products and reactants. If the reaction is only carried out to low conversion we have:

$$\frac{d[CCl_3CH_2CH_2Br]}{d[C_2Cl_6]} = \frac{k_2[CH_2{=}CH_2]}{(\Phi I_a k_6)^{1/2}}$$

Integration of the above expression gives:

$$\frac{[CCl_3CH_2CH_2Br]_f}{[C_2Cl_6]_f} = \frac{k_2[CH_2{=}CH_2]_i}{(\Phi I_a k_6)^{1/2}}$$

where subscript f = final concentration, subscript i = initial concentration.

ΦI_a can be determined by measuring the rate of formation of products against time, so that the rate of the addition step (the step of greatest interest to an organic chemist) can be determined relative to the rate of combination of trichloromethyl radicals. If we now take a different substituted ethylene the expressions will be the same, so it is possible to measure the rate of addition of trichloromethyl radicals to the different ethylenes and different ends of unsymmetrical ethylenes in separate experiments without applying a competitive technique, because the combination of trichloromethyl radicals provides an internal standard for each separate system. Furthermore, the combination of radicals is usually believed to involve little or no activation energy, so the variation in the rate of any one reaction with temperature must be due to variation in k_2 for that particular reaction.

Table 6.10. The addition of trichloromethyl radicals to olefins

$$CCl_3 \cdot + R_2C{=}CR_2 \xrightarrow{\ k_2\ } CCl_3CR_2CR_2$$

$\log A_2$ in l. mole^{-1} sec^{-1}; E_2 in kJ mole^{-1}

Addition to CH$_2$			Addition to CHF			Addition to CF$_2$		
	$\log A_2$	E_2		$\log A_2$	E_2		$\log A_2$	E_2
CH$_2$=CH$_2$	6.5	13.4	CFH=CH$_2$	6.3	22.7	CF$_2$=CH$_2$	6.4	34.9
CH$_2$=CHF	6.4	13.9				CF$_2$=CHF	7.3	29.8
CH$_2$=CF$_2$	6.6	19.3	CFH=CF$_2$	7.2	25.6	CF$_2$=CF$_2$	8.0	25.6

In the case of chlorination we indicated that all the reactions were measured relative to the rate of chlorine atoms with hydrogen. In the addition of trichloromethyl radicals to substituted ethylenes we have indicated how the rates of reaction may all be related to the rate of combination of trichloromethyl radicals. Exactly the same technique has been employed in the examination of hydrogen abstraction reactions of methane where they have all been related to the rate of combination of methyl radicals to yield ethane. It would clearly be desirable to put these reactions on to an absolute scale, and the most usual method for measuring the absolute rate of a radical chain reaction is the rotating sector method.

When reactants like bromotrichloromethane and ethylene are mixed, no reaction occurs until the reaction mixture is illuminated. The rate of the reaction then builds up until the rate of termination equals the rate of initiation and steady-state conditions are established. If the light is now switched off the reaction will continue until all the chains have terminated through the termination step (in the case we are discussing, the combination of the trichloromethyl radicals). However, the light could be turned on again before all the chains had terminated, and in fact, if the dark period was very brief, few chains would be lost. In the present case the initiation rate and hence the radical concentration is proportional to the square-root of the light intensity. If the reaction vessel is illuminated by an intermittent source such that the length of an illuminated period is one third the length of the dark period, we have for very fast intermittency:

$$\text{Rate} = kI^{1/2} \qquad (I = \text{intensity of radiation})$$

i.e. the same as if the system were illuminated by a source of one quarter the intensity; and hence,

$$\frac{\text{Rate(intermittent)}}{\text{Rate(steady)}} = \frac{k(\frac{1}{4}I)^{1/2}}{k(I)^{1/2}} = 0.5$$

and when the intermittency is very slow the effect is the same as if the reaction were only illuminated for one quarter the time; i.e.

$$\frac{\text{Rate(intermittent)}}{\text{Rate(steady)}} = \frac{k\frac{1}{4}(I)^{1/2}}{k(I)^{1/2}} = 0.25$$

Thus, as the intermittency varies from very fast to very slow the rate ratio (intermittent : steady) changes from 0.5 to 0.25 in a way that depends on k. It is possible to draw curves for theoretical values of k for the rate ratio plotted against sector speed, and then to compare the experimental values with the theoretical curve. This technique gives good results provided that there are no alternative chain-termination processes (even though they make only a small contribution).

The Study of Substituents and their Effect on Reaction Rates and Mechanisms

A great deal of the study of organic chemistry is devoted to an investigation of the effects substituents in molecules have on reaction rates and mechanisms. We have seen and discussed many examples of this. For example, nitrobenzene is nitrated much more slowly than benzene and the second nitro group comes in at the *meta*-position. On the other hand, anisole is nitrated much more rapidly than benzene and the new nitro group enters either in the *ortho*- or *para*-positions. The whole of Chapter 8 of Part 2 was devoted to this problem. We saw that the rate of a chemical reaction depends on the free energy change between the reactants and the activated complex, and we argued that, for reactions occurring in solution, substituents which spread the developing charge reduce the free energy of the activated complex and therefore accelerate the reaction. Where substituents concentrate the developing charge they raise the free energy of the activated complex and therefore retard the reaction. We have also seen that a nitro group which is both electron-attracting and electron-accepting has a retarding influence not only on nitration but also on halogenation, sulphonation, and any other electrophilic addition-with-elimination reactions. In order to obtain a quantitative measure of the activating and deactivating effect of groups, we could measure the rate of nitration, say, at the *para*-position in anisole and compare it with the rate of attack on benzene. We can do this most accurately by doing competitive studies in which both aromatic compounds are present together, and if one is much more reactive than the other we can compensate this by having the two compounds present in vastly different concentrations.

By this means so-called *partial rate factors*, by which the rate of attack at the position in question is compared with the rate of attack at one position in benzene, have been measured. The calculation of partial rate factors is clarified by the following example.

In the nitration of toluene by nitric acid in acetic anhydride at 25° 33.9% of the *para*- and 2.8% of the *meta*-isomer was obtained, and the relative reactivity of toluene to benzene was 23 (these are percentages of the total nitrotoluene). The reactivity of the *para*-position of toluene relative to benzene is clearly

$$\frac{33.9}{100} \times 23$$

and since there are six positions available for nitration in benzene and only one *para*-position in toluene, the reactivity of the latter relative to a single position in benzene is

$$6 \times \frac{33.9}{100} \times 23 = 46.8$$

Likewise, the partial rate factor for the *meta*-position in toluene is

$$\frac{6}{2} \times \frac{2.8}{100} \times 23 = 1.93.$$

A partial rate factor greater than 1 for a given position indicates that the group in question is activating that position for the given reaction.

Different reagents show different selectivities, and a selectivity factor has been defined as the logarithm of the partial rate factor for the *para*-position over the partial rate factor for the *meta*-position for the reaction of toluene:

$$S_f = \log(f_p \text{ toluene}/f_m \text{ toluene})$$

The partial rate factors for some common electrophilic addition-with-elimination reactions with toluene are illustrated (along with S_f for each reaction).

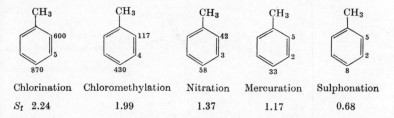

Chlorination	Chloromethylation	Nitration	Mercuration	Sulphonation
S_f 2.24	1.99	1.37	1.17	0.68

Notice that for some reactions the partial rate factor for attack at the *ortho*-position is appreciably lower than the partial rate factor for attack at the *para*-position. This is particularly noticeable when the reagent is likely to be of a bulky nature, for instance in chloromethylation and mercuration, and there is every reason to believe that this is simply the result of the substituent methyl group getting in the way of the attacking electrophile. A very good linear relation is found between the logarithm of the partial rate factor for toluene and the selectivity factor. This is equivalent to a linear relationship between logarithm of the partial rate factor for the *para*-position against the logarithm of the partial rate factor for the *meta*-position, and the fact that it occurs strongly suggests that all these electrophilic addition-with-elimination reactions have very similar activated complexes.

Much thought and effort has gone into comparing the effects of substituents in one reaction with the effects of the same substituent on the course of a different reaction. We would hardly be surprised if we found that there was a linear relation between the logarithms of the rates of hydrolysis of a series of nuclear substituted ethyl benzoates and the logarithms of the rates of hydrolysis of the corresponding methyl benzoates. Or, taking the argument one stage further, instead of comparing the rate of hydrolysis of methyl esters with the rate for ethyl esters it would be of interest to compare the rates of hydrolysis of both sets of esters with the dissociation constants of the corresponding benzoic acids.

A linear relation is found between the logarithm of the rate of ester hydrolysis and the logarithm of the dissociation constant of the corresponding acid, and is no great matter for surprise but something

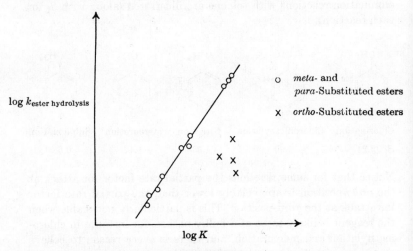

Figure 6.12

we would expect. In practice we find a good linear relation for *meta-* and *para*-substituted benzoic acids but no correlation at all between the dissociation constant and the rate of ester hydrolysis for *ortho*-substituted benzoic acids or aliphatic acids (see Figure 6.12). The line we obtain obeys the equation

$$\log k = \rho \log K + A$$

where k is the rate constant of the hydrolysis, K is the ionization constant of the corresponding benzoic acid, ρ (rho) is the slope, and A the intercept. Similar linear relationships are found to apply to the rate and equilibrium constants of many of the side-chain reactions of benzene derivatives. The equation, of course, holds where the substituent is hydrogen, so if we denote the rate constant and equilibrium constant of the unsubstituted benzoic acid by k^0 and K^0 respectively, we have

$$\log k^0 = \rho \log K^0 + A$$

Hence, by subtraction,

$$\log k - \log k^0 = \rho(\log K - \log K^0)$$

Hammett suggested calling the logarithm of the ratio of the dissociation constants of the unsubstituted and substituted acids σ (sigma); i.e.

$$\sigma \equiv \log K - \log K^0$$

The original equation now becomes

$$\log k - \log k^0 = \rho\sigma$$

This expression is usually known as the Hammett equation. It is apparent that σ depends only on the nature of the substituent and is designated the substituent constant, while ρ, which is the gradient of the linear plot, varies with the nature of the reactions but is independent of the nature of the substituent and is termed the reaction constant. Notice that the equation can only be applied to *meta-* and *para-*substituted compounds. No such relation exists for *ortho-*substituted compounds.

It is now possible to establish a scale of polar effects for *meta-* and *para-*substituents derived from the ionization constants of benzoic acids. Typical values are given in Table 6.11. σ represents the ability

Table 6.11.

X $\diagdown$ $\langle ⎔ \rangle$ —CO_2H	Hammett σ-values	
X	*meta*	*para*
H	0	0
CH$_3$	−0.07	−0.17
NH$_2$	−0.16	−0.66
OCH$_3$	+0.12	−0.27
Cl	+0.37	+0.23
CN	+0.56	+0.66
CF$_3$	+0.43	+0.55
NO$_2$	+0.71	+0.78
$^+$N(CH$_3$)$_3$	+0.91	+0.86
SO$_2$CH$_3$	+0.60	+0.72

Table 6.12

Reaction			ρ
$X\text{---}C_6H_4\text{---}CO_2H$	$\rightleftarrows$	$X\text{---}C_6H_4\text{---}CO_2^- + H^+$ (water, 25°)	+1.00
$X\text{---}C_6H_4\text{---}OH$	$\rightleftarrows$	$X\text{---}C_6H_4\text{---}O^- + H^+$ (water, 25°)	+2.20
$X\text{---}C_6H_4\text{---}CH_2CO_2H$	$\rightleftarrows$	$X\text{---}C_6H_4\text{---}CH_2CO_2^- + H^+$ (water, 25°)	+0.49
$X\text{---}C_6H_4\text{---}CH_2CH_2CO_2H$	$\rightleftarrows$	$X\text{---}C_6H_4\text{---}CH_2CH_2CO_2^- + H^+$ (water, 25°)	+0.21
$X\text{---}C_6H_4\text{---}COOC_2H_5 + H^+$	$\rightarrow$	$X\text{---}C_6H_4\text{---}COOH + C_2H_5OH$ (60% ethanol, 100°)	+0.03
$X\text{---}C_6H_4\text{---}COOC_2H_5 + \bar{O}H$	$\rightarrow$	$X\text{---}C_6H_4\text{---}CO_2^- + C_2H_5OH$ (60% acetone, 0°)	+2.50
$X\text{---}C_6H_4\text{---}CONH_2 + H^+$	$\rightarrow$	$X\text{---}C_6H_4\text{---}COOH + NH_4^+$ (water, 100°)	+0.12
$X\text{---}C_6H_4\text{---}CH\text{=}CHCO_2H$	$\rightleftarrows$	$X\text{---}C_6H_4\text{---}CH\text{=}CHCO_2^- + H^+$ (water, 25°)	+0.47

$$X\text{-}C_6H_4\text{-}CH=CHCOOC_2H_5 + \bar{O}H \longrightarrow X\text{-}C_6H_4\text{-}CH=CHCO_2^- + C_2H_5OH \qquad +1.33 \quad (\text{water, } 25°, \text{ pH } 12.6)$$

$$X\text{-}C_6H_4\text{-}CHO + KMnO_4 \longrightarrow X\text{-}C_6H_4\text{-}COOH \qquad +1.80 \quad (\text{ethanol, } 20°)$$

$$X\text{-}C_6H_4\text{-}CHO + CN^- + H^+ \longrightarrow X\text{-}C_6H_4\text{-}CH \begin{smallmatrix} OH \\ CN \end{smallmatrix} \qquad +2.33 \quad (\text{water, } 25°)$$

$$X\text{-}C_6H_4\text{-}\overset{+}{N}H_3 \longrightarrow X\text{-}C_6H_4\text{-}NH_2 + H^{++} \qquad +2.77 \quad (\text{50\% acetone, } 60°)$$

$$X\text{-}C_6H_4\text{-}CH_2Cl + \bar{O}H \longrightarrow X\text{-}C_6H_4\text{-}CH_2OH + Cl^- \qquad -1.69 \quad (\text{ethanol, } 25°)$$

$$X\text{-}C_6H_4\text{-}O^- + C_2H_5I \longrightarrow X\text{-}C_6H_4\text{-}OC_2H_5 + I^- \qquad -0.99 \quad (SO_2, 0°)$$

$$X\text{-}C_6H_4\text{-}CCl_3 \longrightarrow X\text{-}C_6H_4\text{-}\overset{+}{C}()_3 + Cl^- \qquad -3.97 \quad (SO_2, 0°)$$

$$X\text{-}C_6H_4 + HNO_3 \longrightarrow X\text{-}C_6H_4\text{-}NO_2 \qquad -6.0 \quad (\text{nitromethane, } 25°)$$

of the substituted aromatic derivative to accept or donate electrons; in other words, it measures the effect of X on C* whereas ρ indicates

the extent to which the aromatic compound is behaving as a nucleophile or an electrophile in the process concerned; i.e. it is a measure of the susceptibility of the main reaction group A* to changes at C*.

To obtain ρ for a given reaction series it is necessary to measure rate or equilibrium constants for a number of compounds, each having the reaction centre under consideration and each having a different ring substituent with a known σ value. The logarithms of the measured constants are plotted against the corresponding σ values and the slope of the best straight line through the points on such a plot is the ρ value for the reaction series at hand.

A few reaction constants for both rates and equilibria are given in Table 6.12.

It is obvious from the picture which we have presented that major changes in the magnitude of the value of ρ will be brought about by differences in the ease of transmission of polar effects from the point of substitution (C*) to the point of reaction (A*); for example, the insulating effect of a single bond will make a polar effect less easy to transmit to the point of reaction. Thus the effect of substituents on the ionization of phenols is greater than their effect on the ionization of benzoic acids, because of the direct resonance between $-O^-$ in the phenolate anion and electron-withdrawing substituents in the *para*-position. In fact, a new set of σ values (σ^- parameter) has been defined to correlate such reactions, but we will deal with these later.

Similarly, double bonds transmit polar effects better than single bonds, so it is no surprise to find that the ρ value for the ionization of cinnamic acids is greater than for β-phenylpropionic acids.

Since these ionization processes we have just discussed all involve the production of a negatively charged species (RCO_2^-, RO^-), electron-withdrawing substituents (σ positive) will increase the equilibrium constant, and hence ρ is positive. Provided that ρ is numerically substantial (i.e. >0.5) its sign indicates the sign of the charge on the transition state; i.e. a positive ρ value corresponds to a reaction in which the transition state is more negative than the reactant while reactions in which the transition state is more positive than the reactant have negative ρ values.

For example, consider the following reaction:

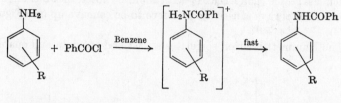

$\rho = -2.78$ (at 25°)

This value for ρ shows that the first step is the rate-determining step. If the second step were rate-determining the transition state would be less positive than the reactant ($Ar\overset{+}{N}H_2COPh$), and ρ would therefore be positive.

It is also important to realize that when the mechanism of a reaction changes because of the presence of certain substituents, curvature in the $\rho\sigma$ relationship can occur. A striking example of an abrupt change in mechanism accompanied by an abrupt change in slope of the $\rho\sigma$ plot is the hydrolysis of benzoate esters in 99.9% sulphuric acid at 45°. Hydrolysis of the methyl esters obeys a $\rho\sigma$ relationship with good precision, the rate decreasing consistently as the substituents are made more electron-attracting, i.e. ρ is negative. This is to be expected for the acyl–oxygen fission mechanism:

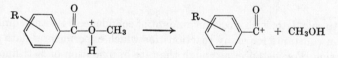

But for the ethyl esters it appears that strongly electron-attracting groups cause a sudden shift to alkyl–oxygen fission:

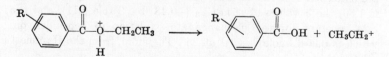

and so accelerate the reaction (i.e. ρ becomes positive).

It is clear in these examples that the more stable $CH_3CH_2^+$ carbonium ion permits the formation of a less positive, and therefore more stable, transition state when the benzene ring contains electron-attracting groups. This is not possible for methyl esters, CH_3^+ being much less stable than $CH_3CH_2^+$. It should be noted that a change in mechanism always causes the $\rho\sigma$ curve to be concave up (see Figure 6.13).

The Hammett relation correlates a vast body of data and enables

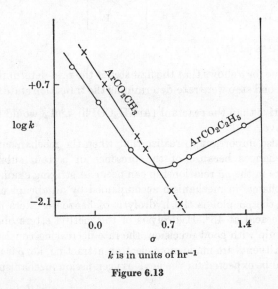

k is in units of hr^{-1}

Figure 6.13

us to calculate, with a certain amount of confidence, rate and equilibrium constants for many reactions for which experimental data are lacking. For instance, although the rate of oxidation of m-trifluoromethylbenzaldehyde by aqueous permanganate has not been measured, its rate can be estimated from the ρ value for permanganate oxidation of aromatic aldehydes, which from Table 6.12 we see is 1.80, the σ values of $meta$-CF_3 (+0.43; see Table 6.11), and the rate constant for the permanganate oxidation of benzaldehyde, which is 0.90 l. mole^{-1} sec^{-1} at 25°. Therefore, from the Hammett relationship:

$$\log k/0.90 = 0.43 \times 1.80$$

and

$$k = 5.3 \, \text{l. mole}^{-1} \text{sec}^{-1}$$

Correlations have also been found between $\rho\sigma$ and the logarithm of the chemical shifts of ^{19}F NMR absorptions in substituted fluorobenzenes and in the intensity of the infrared absorption of carbonyl groups, cyano groups, and diazo groups attached to substituted benzene rings. Although the success of the relationship has led many people to believe that the equation has a more fundamental basis than it really has, our whole picture of substituent effects really requires a relationship of this kind. For example, we would expect the donating power of a methoxy group or the accepting power of a nitro group to depend on the nature of the process we are concerned with. In fact this is well brought out in the correlations obtained with the Hammett equations.

We have already seen (page 324) that the ionization of phenols requires a slightly different set of σ values from that derived from K_a data for benzoic acids. This is particularly true for *para*-substituents capable of stabilizing a negative charge by resonance. Use of these new σ values, denoted as σ^- values, improves the linearity of the Hammett correlation for phenols, phenolic esters, anilines, and anilides, and is particularly useful for correlating aromatic nucleophilic substitution.

Just as direct resonance with a negative centre requires new σ values, particularly for electron-withdrawing substituents, so resonance with a positive centre will mean amending the σ values, particularly for electron-releasing substituents. This new series of parameters, denoted as σ^+ values, are computed from an equation similar to the Hammett $\rho\sigma$ relation in which the S_N1 hydrolysis of substituted benzyl chlorides is used as the standard reaction in place of the dissociation constants of corresponding benzoic acids.

Table 6.13. Some typical σ constants compared with σ^+ constants

	σ	σ^+		σ	σ^+
p-CH$_3$O	-0.27	-0.78	p-CH$_3$	-0.17	-0.31
m-CH$_3$O	-0.12	$+0.05$	m-NO$_2$	$+0.71$	$+0.67$
m-CH$_3$	-0.07	-0.07	p-NO$_2$	$+0.78$	$+0.79$

Typical ρ values for the $\rho\sigma^+$ correlation are given in Table 6.14. The values are always negative and their magnitudes are often large, in agreement with our expectation that these reactions are facilitated by electron-donating substituents. The resemblance of the various reactions to the σ^+ standard reaction can also be seen from the following formulae which represent either transition states or products of the rate-determining steps.

Table 6.14

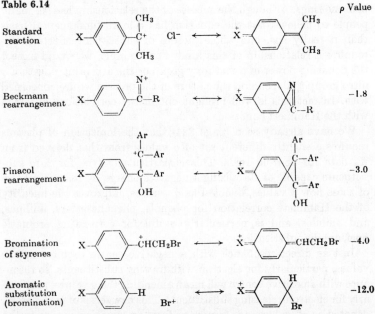

		ρ Value
Standard reaction		
Beckmann rearrangement		-1.8
Pinacol rearrangement		-3.0
Bromination of styrenes		-4.0
Aromatic substitution (bromination)		-12.0

The tremendous success of the Hammett equation in correlating a vast range of data has naturally encouraged chemists to seek a similar correlation for aliphatic compounds. The only such correlation to prove at all successful is that due to Taft.

No correlation is observed if a simple relation of the Hammett type is employed. However, a rather good value for the polar substituent constant, σ^*, may be obtained by selecting an ester having that substituent α to the carbonyl group, and then comparing the specific rates for acidic and basic hydrolysis of that ester. Taft then defined σ^* as:

$$\sigma^* = \frac{1}{2.5}\left[\log\left(\frac{k}{k_0}\right)_{\text{B}} - \log\left(\frac{k}{k_0}\right)_{\text{A}}\right]$$

where $(k/k_0)_B$ is the ratio of the rates of base ester hydrolysis and $(k/k_0)_A$ is the ratio of the rates of the acid-catalysed hydrolysis. The numerical constant $1/2.5$ is arbitrarily introduced to bring the σ^* aliphatic constants into line with the σ constants of the Hammett equation. The basis of this definition is as follows. In the case of *meta-* and *para-*substituted benzene derivatives, alkaline hydrolysis of ethyl benzoate shows a large ρ factor ($+2.50$) whereas the ρ value for the acid hydrolysis of ethyl benzoate is very small indeed ($+0.03$). This means that alkaline hydrolysis is greatly affected by polar substituents whereas acid hydrolysis is not. On the other hand, the rates of acid hydrolysis of aliphatic esters are strongly affected by substituents. Since there is no reason why these esters should be more susceptible to polar effects than are aromatic esters, and since resonance effects cannot be transmitted along aliphatic chains, we may conclude that steric effects are intervening in the aliphatic series. With negligible polar and resonance effects, the ratio $\log(k/k_0)_A$ becomes a measure of the steric effect of the substituent. Taft assumes that because the activated complex in a normal acid-catalysed hydrolysis differs from that in base hydrolysis only by the presence of two additional protons, steric effects in the two types of hydrolyses

$$\left[\begin{array}{c} O \\ \| \\ R-C\cdots\cdots O-R' \\ | \\ H \\ | \\ O \\ / \quad \backslash \\ H \qquad H \end{array}\right]^{+} \qquad \left[\begin{array}{c} O \\ \| \\ R-C\cdots\cdots O-R' \\ | \\ OH \end{array}\right]^{-}$$

should be virtually the same. Since both polar and steric effects are important in base hydrolysis, the term $\log(k/k_0)_B$ is a measure of steric plus polar effects. The difference $[\log(k/k_0)_B - \log(k/k_0)_A]$ is therefore taken as a measure of the polar effect alone. By using σ^* then, we get a correlation analogous to the Hammett one, namely:

$$\log k - \log k^0 = \rho^* \sigma^*$$

Some typical σ^* values are shown in Table 6.15.

The Taft equation is really a much more arbitrary correlation than

Table 6.15. Typical σ^* constants for aliphatic compounds

R	σ^*	R	σ^*	R	σ^*
CCl_3	$+2.7$	$CClH_2$	$+1.1$	$C_6H_5CH_2$	$+0.2$
CCl_2H	$+1.9$	C_6H_5	$+0.6$	CH_3	0.0
CH_3CO	$+1.7$	H	$+0.5$	C_2H_5	-0.1
				$t\text{-}C_4H_9$	-0.3

the simple Hammett equation, and far fewer applications have been found. Nonetheless correlations of this kind are useful. Two interesting examples are shown below.

For the ionization of various acids, RCO_2H, in water at $25°$, $\rho^* = 1.72$; i.e.

$$\log K/K_0 = 1.72\sigma^*$$

For the catalysis of dehydration of $CH_3CH(OH)_2$ by RCO_2H, $\rho^* = 0.80$; i.e.

$$\log k/k_0 = 0.80\sigma^*$$

Therefore,

$$\frac{\log(k/k_0)}{\log(K/K_0)} = \frac{0.80}{1.72} \approx 0.5$$

This is, of course, equal to the value of α obtained by plotting $\log k$ against $\log K$ for the dehydration of acetaldehyde hydrate by various acids (see page 293).

Many other modifications of the Hammett equation have been proposed but the proliferation of such terms undermines the whole value of the relationship. Perhaps far too much has been read into the Hammett equation. It simply represents the comparison of one set of free energy data with another set. The value of the relation lies in the way it provides a basis of comparison for a very wide variety of chemical processes. The very fact that a reaction shows a correlation of the Hammett $\rho\sigma$ type suggests that the activated complex has considerable polar character. There is no immediate reason for supposing that radical reactions involve polar transition states but the rates of atomic bromination of *meta*- and *para*-substituted toluenes show a good correlation with Hammett σ (or better with σ^+ constants), and this provides strong evidence that the activated complex has considerable polar character. Conversely, the failure of reactions to show correlations with the Hammett equation has on occasion been used to detect the change in a reaction mechanism as the substituents are changed.

Classical and Non-classical Ions

We have already seen that the most widely investigated effects of a substituent in an organic molecule, on the reactions of that molecule, are electronic effects (inductive, conjugative) transmitted through the carbon skeleton, and steric effects. In addition, however, some substituents influence a reaction by stabilizing the transition state by becoming bonded or partially bonded to the reaction centre. This behaviour is called neighbouring group participation.

There are two principal types of evidence that point to neighbouring group participation. Firstly, if such participation occurs during the rate-determining step, thus stabilizing the transition state, the reaction is almost certain to be significantly more rapid than other reactions that are similar but do not involve such participation. If such an increased reaction rate results, the neighbouring group is then said to provide anchimeric assistance. This phenomenon is illustrated by the relative rates of acetolysis of the compounds in Table 6.16. As well as facilitating the slow step, the neighbouring

Table 6.16

	Relative rate of acetolysis
$CH_3[CH_2]_3OBs$	1
$CH_3O[CH_2]_3OBs$	0.63
$CH_3O[CH_2]_4OBs \equiv$	657

$$\left(Bs \equiv -SO_2-\!\!\!\!\!\bigcirc\!\!\!\!\!-Br \right)$$

group may be involved in the formation of a discrete intermediate (a dip in the energy profile diagram) not involved in the unassisted reaction. The resultant intermediate may also react to yield a product which would not be expected in the absence of participation; this may

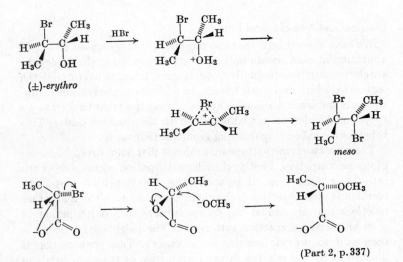

(±)-*erythro*

meso

(Part 2, p. 337)

be a product with retained configuration or one in which a ring is formed.

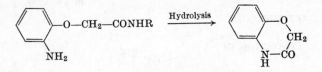

Many atoms and groups are known to participate as neighbouring groups. Generally speaking the most effective groups are large and polarizable and are those which would generate nucleophilic anions if released, e.g.

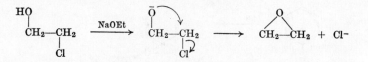

In general, it has been found that intramolecular displacements leading to epoxides are, as a group, thousands of times faster (under comparable conditions) than their intermolecular counterparts, the attack of alkoxides on alkyl chlorides. This large preference for intramolecular reaction, even at the expense of forming a strained ring, is due to probability factors. In collision-theory terms, the nucleophile is permanently held next to the carbon it must attack, so reaction can occur whenever the species picks up enough energy. In transition-state terms, an ordinary S_N2 reaction involves a loss of entropy when the nucleophile and substrate are tied down in the transition state. In an internal reaction there is no need to tie down a second reactant, so the entropy of activation is much more favourable. This can even make up for the unfavourable enthalpy associated with making a strained ring (Chapter 3, p. 193).

The neighbouring groups we have considered so far are nucleophilic in the classical sense, since each has unpaired electrons available for coordination or displacement. The aromatic ring also has nucleophilic character, so we might expect that under favourable conditions neighbouring group participation by aryl groups could be observed.

One such situation appears to occur in the case of β,β,β-triphenylethyl chloride which solvolyses in formic acid at a rate 60,000 times that of neopentyl chloride. Phenyl groups attract electrons inductively, so they would destabilize the already unstable primary cation or a transition state which resembled the unrearranged cation, but if migration is simultaneous with ionization the rate enhancement can be understood.

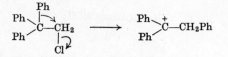

Even more unusual examples are known. For instance, the acetolysis (solvolysis in acetic acid) of optically active *threo*-3-phenyl-2-butyl toluenesulphonate affords the racemic *threo*-acetate.

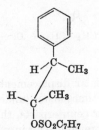
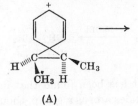

threo-3-Phenyl-2-butyl
toluenesulphonate

(A)

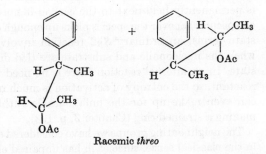

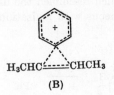

Racemic *threo*

The formation of only the *threo*-acetate shows that the toluene-sulphonate group is replaced with retention of configuration, as in other examples of neighbouring group participation. The formation of completely racemized product is expected from the intermediate carbonium ion (A) which has a plane of symmetry. Instead of writing all the contributing structures for this ion it can be represented as (B).

H₃CHC⌐----⌐CHCH₃

(B)

An ion of this kind has been called a phenonium ion—an example of a non-classical carbonium ion. If phenyl participation had not occurred the products would have been diastereoisomers, not enantiomers.

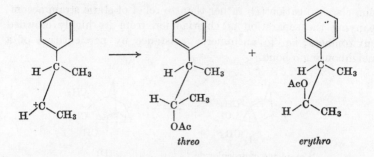

threo   *erythro*

There is a certain amount of other evidence for the existence of phenonium ions; e.g. in favourable instances intermediate dienones have been isolated.

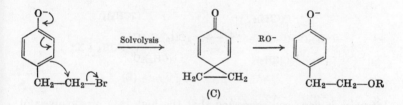

(C)

In this case the formation and decomposition of the dienone (C) have been followed spectrophotometrically, and by the use of highly acidic media (e.g. FSO_3H–SO_2–SbF_5) bridged ions such as those illustrated below have been directly observed by NMR spectroscopy.

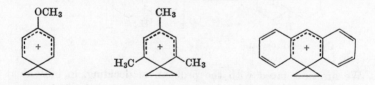

The non-classical ion concept is now very fashionable, and there are many cases for which non-classical structures have been considered. Thus, the fact that camphene hydrochloride undergoes ethanolysis at a rate 6000 times greater than t-butyl chloride has been attributed to the smaller activation energy for the formation of a non-classical cation (D) rather than to relief of steric strain accompanying the separation of chloride ion from its highly crowded environment, i.e. to anchimeric assistance by participation of a neighbouring σ bond.

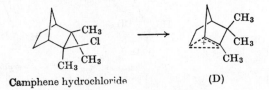

Camphene hydrochloride   (D)

Similarly, it has been suggested that the formation of a methyl bridged tri-t-butylcarbinyl cation (E) might provide the driving force for the observed fast solvolysis of tri-t-butylcarbinyl derivatives.

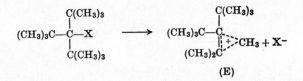

(E)

Likewise it has been suggested that the high rate of solvolysis of cyclodecyl tosylate might be due to the formation of a stabilized bridged cyclodecyl cation

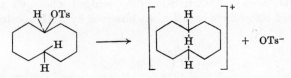

Cyclodecyl tosylate

We are now faced with the problem of deciding, in individual systems, whether the structure of the intermediate should be represented as a classical form, or as an equilibrating pair (or set) of cations, or as a resonance hybrid (i.e. a non-classical ion) so stable that these prior alternative structures need not be considered. Thus the norbornyl cation may be represented by one or all of the following structures.

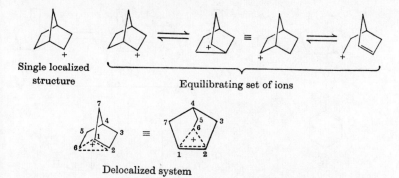

Single localized
structure

Equilibrating set of ions

Delocalized system

For ions, such as the phenethyl cation in its bridged form, which possess sufficient electron-pairs for all the required bonds, it requires no extension of generally accepted bonding concepts to account for these structures (cf. intermediates in normal electrophilic aromatic substitution). However, ions such as the norbornyl cation do not possess, in their σ-bridged form, sufficient electrons to provide a pair for all the bonds required by the proposed structures, and a new bonding concept, not yet definitely established in carbon structures, is required.

The problem can, of course, be considered as a logical extension of neighbouring group participation from nucleophilic groups, to π-groups and finally to σ-groups. It would obviously be highly desirable to have a clear-cut case of σ-participation, and the norbornyl system has been subjected to a great deal of careful study to ascertain whether it, in fact, provides that case.

Consider the solvolysis of *exo*-bicyclo[2,2,1]hept-2-yl bromobenzenesulphonate* in acetic acid. The observed result of the acetolysis of this optically active material is the formation of completely racemic *exo*-acetate, i.e. an equimolar mixture of (D) and (E). One explanation of this result would be as illustrated, where a simple carbonium ion (A) is formed initially. This then rearranges to its mirror image, ion (B), and the resulting racemic mixture of carbonium ions can then react with acetic acid, exclusively on the unhindered underside of the ring, to yield the racemic mixtures of products. The

---

* See p. 150. '*exo*' indicates the less hindered configuration of the substituent. The more hindered epimer is called '*endo*' (see p. 339).

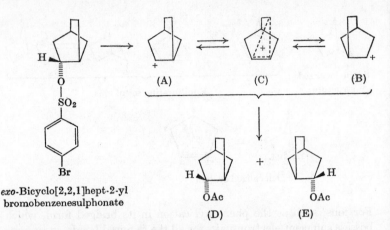

exo-Bicyclo[2,2,1]hept-2-yl
bromobenzenesulphonate

alternative possibility is that solvolysis leads directly to a symmetrical bridged ion (C) more stable than either of the open ions.
Reaction of (C) with acetic acid would then produce the racemic
mixture of acetates without involvement of the classical ions (A) and
(B).

Support for the formation of such a symmetrical ion as the initially
formed intermediate in the solvolysis of norbornyl arenesulphonates
such as (F) was demonstrated in the following way.

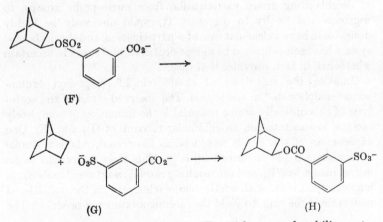

The optically active *exo*-ester (F) with a nucleophilic *meta*-
carboxylate ion was used in order that the product, the carboxylic

ester (H), would be formed with only the slightest molecular movement, after ionization. Consequently, if an unsymmetrical intermediate of an appreciable life, such as (G), were formed, it would be expected to yield optically active product. The product however was wholly racemic.

Another piece of evidence supporting the proposal of a non-classical structure for the norbornyl cation and its substitution products is as follows. Acetolysis of *exo*-norbornyl brosylate is 350 times as fast as acetolysis of the corresponding *endo*-derivative. The high rate for the *exo*-compound is attributed to σ-participation (cf. neighbouring group participation) by the 1,6-bonding pair; it is

exo-Norbornyl brosylate

endo-Norbornyl brosylate

considered that the *endo*-derivative is stereochemically unsuited for such participation. Furthermore, solvolysis of the *endo*-isomer gives virtually wholly racemic *exo*-product. One school of thought finds it difficult to explain this behaviour without postulating the intermediacy of a non-classical ion.

On the other hand, attempts have been made to account for these observations without utilizing the formation of a bridged cation.

For example, the fact that the *exo*-compound solvolyses faster than the *endo*-compound may not necessarily mean that the *exo*-derivative exhibits an enhanced rate. It may mean that the rate of the *exo*-derivative is normal and that of the *endo*-compound is unusually slow because of the possibility of steric hindrance to ionization and steric hindrance to substitution in the *endo*-compound. Thus it has been suggested that the behaviour of these compounds in solvolysis simply conforms to the general pattern of reactivity observed in norbornyl derivatives not involving carbonium ions. In other words, all the observed rate enhancements etc. can be accounted for in terms of relief of steric strain. Thus, much of the general chemistry of norbornane is dominated by the far greater steric accessibility of the *exo* as compared to the *endo* position. For example, in many free-radical substitution reactions the *exo*-product is produced preferentially.

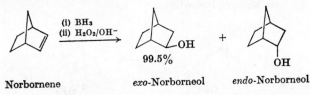

Similarly, many additions to norbornene proceed predominantly from the *exo* direction; thus the hydroboration–oxidation of norbornene produces 99.5% of *exo*-norborneol, with only 0.5% of the *endo*-isomer.

It has also been reported that *exo*-norbornyl derivatives undergo *cis* E2 elimination, whereas *endo*-norbornyl derivatives undergo *trans* E2 elimination. In each case it is the *exo*-hydrogen that participates in the elimination. Thus, in this system the elimination appears to be controlled primarily by the greater steric availability of the *exo*-hydrogen rather than by any stereoelectronic preference of the reaction mechanisms for either *cis*- or *trans*-elimination.

Such arguments reveal that there may be a reasonable alternative to non-classical structures in accounting for the behaviour of bicyclic carbonium ions.

So the big question still is "whether non-classical ions are reaction intermediates or merely transition states for the interconversion of classical ions". Actually, recent Raman spectral studies suggest that under certain conditions the most stable structure of the norbornyl cationic intermediate is the edge-protonated cyclopropane (A).

(A)

Finally, it is probably fair to say that the present state of our knowledge points to bridged and open carbonium ion theory complementing one another. Either one alone fails to explain a great body of experimental data.

Non-kinetic Methods

In discussing the tools used for the investigation of reaction mechanisms we have postponed until last the use of non-kinetic methods.

A prerequisite to any meaningful study of reaction mechanisms is a knowledge of the structures of starting materials and products. The mechanism of a reaction must account for the products. Thus, when chlorine is added to cyclohexene, the product is *trans*-1,2-dichlorocyclohexene and not the *cis*-isomer (Chapter 3, p. 176). Clearly, any mechanism for this reaction must account for the *trans*-addition. Likewise we soon become aware, because of the nature of the products, that the bromination mechanism which operates when toluene is illuminated in the vapour phase in the presence of bromine is different from that when the bromination is carried out in the liquid phase in the presence of a halogen carrier. Note that in these examples we have suggested no mechanistic details; nevertheless, the correct identification of the products strongly suggests differences in mechanism.

We must of course consider not only the main products of a reaction but also the by-products. The by-products are not always formed in

12

independent reactions but are sometimes predictable in terms of mechanisms, the first steps of which are identical with those of the main reaction. For example, the formolysis of cyclooctene oxide gives the expected *trans*-cyclooctane-1,2-diol along with substantial amounts of *cis*-cyclooctane-1,4-diol (Chapter 3, p. 187).

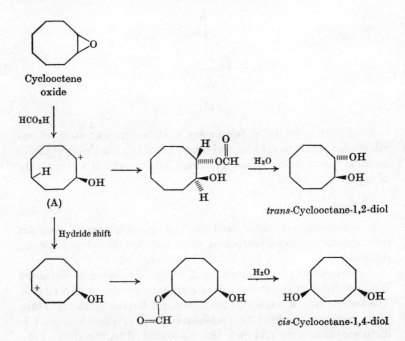

A mechanism that has been suggested for this reaction explains the by-product as resulting from a transannular shift of hydride (hydrogen with its electron pair) to the initially formed carbonium ion (A). Although the distance seems large in the above structures, models show that the migrating hydrogen can actually be quite close to the positive carbon, and the fact that both *trans*-cyclooctane-1,2-diol and *cis*-cyclooctane-1,4-diol are formed argues strongly for the initial formation of the carbonium ion (A).

Before we can make any attempt to describe the complete course of a chemical reaction we must also be familiar with the structures of the intermediate compounds, or ions, which form and decay as the reaction proceeds. Such information can often be obtained by either trapping the intermediate or by detecting it by some suitable technique.

Sometimes an intermediate which cannot be isolated may be detected by adding to the reaction a 'trapping agent'. This is added in order to combine with the intermediate to form a product that we would be unable to account for otherwise. Thus the addition of bromine to many olefins in polar solvents is thought to proceed

through an intermediate bromonium ion  . Strong

evidence for such an intermediate is that it can be diverted from its usual reaction course by the presence of nucleophilic reagents other than Br⁻. For example, when bromine is added to stilbene in methanol, the bromo-ether (B) is also isolated from the reaction mixture. This is presumably formed by the intermediate (A) reacting with methanol instead of with Br⁻.

$C_6H_5CH{=}CHC_6H_5$

  Stilbene

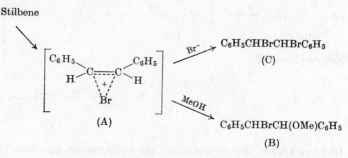

We must note that isolation of the ether (B) is not sufficient proof that the bromination proceeds through intermediate (A). For example, it must be demonstrated that the dibromide (C) does not readily react with methanol to yield the bromo-ether. It must also be shown that methanol and bromine alone do not react to give an intermediate which itself reacts with the olefin to give the bromo-ether. Similarly, dehydrobenzene, formed e.g. by decomposition of the *o*-carboxybenzenediazonium ion, can be trapped as triptycene by the addition of anthracene to the system.

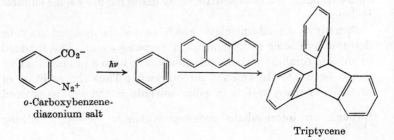

o-Carboxybenzene-
diazonium salt

Triptycene

A variation of the idea of intermediate trapping is the use of a reagent of such a structure that the intermediate, if formed at all, would have peculiar properties. For instance, in the solvolysis of neopentyl bromide the intermediate carbonium ion, if formed, should instantly rearrange to the more stable tertiary carbonium ion.

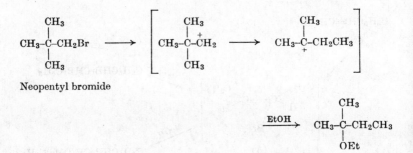

Neopentyl bromide

In accord with this expectation, the substitution products of this reaction in 80% ethanol are t-pentyl ethyl ether and t-pentyl alcohol.

In a few favourable cases the presence of an intermediate may be detected by physical measurements. Thus introduction of the *o*-carboxybenzenediazonium ion into a mass spectrometer yields a simple spectrum with peaks at m/e 28, 44, 76, and 152. These peaks clearly correspond to the fragmentation pattern shown below, thus confirming the intermediacy of dehydrobenzene in the decomposition of that ion.

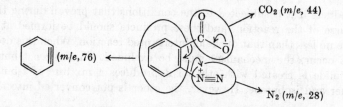

Moreover, the spectrum is time-variable—the peak at m/e 76 disappears rapidly as the peak at m/e 152 grows. The latter peak is obviously due to the formation of biphenylene, arising from the dimerization of dehydrobenzene.

Similarly, free radicals of long life may be detected by magnetic measurements (Part 2, p. 379). Thus ESR spectroscopy has provided evidence for the existence of the relatively transient radical (A)

(A)

detected in a flow system from the interaction of $\cdot OH$ and benzene. The splitting pattern shows that it is not the $\cdot OH$ radical that is being seen.

At this point it is appropriate to mention the important advance of the direct observation of carbonium ions by NMR spectroscopy. Originally it was kinetic and stereochemical evidence that established carbonium ions as intermediates in organic reactions; but until very recently only fragmentary evidence was available for their direct observation in solution. Olah and coworkers have utilized SbF_5 as both the Lewis acid and solvent, and the very strongly acidic solvent system FSO_3H–SO_2–SbF_5. Stable alkyl-carbonium hexafluoroantimonate salts are formed, and extensive NMR study has left no doubt as to their structure; e.g.

$$(CH_3)_3CF + SbF_5 \rightleftharpoons (CH_3)_3C^+SbF_6^-$$

If an intermediate is postulated in a reaction mechanism it is desirable that it is detected either by chemical or physical means. Equally important, such an intermediate must lead to the correct products if it is subjected to the conditions that prevail during the course of the reaction, and these products should be formed at a rate no less than that of the uninterrupted reaction. When this does not occur, the mechanism must be excluded. Thus when t-butyl bromide is treated with ethanol it produces a mixture of t-butyl ether and isobutene. However, the ether is not converted into the

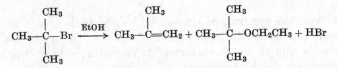

olefin under these conditions, and vice versa. Therefore, the ether cannot be an intermediate in the formation of the olefin.

On the other hand, we can feel quite confident that the anion $RCONBr^-$ is a probable intermediate in the Hofmann rearrangement; e.g.

$$RCONH_2 \xrightarrow{OBr^-} RCONHBr \xrightarrow{OH^-} RCONBr^- \xrightarrow{-Br^-} RN{=}C{=}O \xrightarrow{H_2O} RNH_2 + CO_2$$

for, with care, salts containing this anion may be isolated, and such salts undergo the conversion into isocyanate at a rate no less than that for the brominated amide.

Mechanisms may also be verified by noting the failure of a reaction in cases where the postulated intermediate would be impossible or of very high energy. For example, the Hofmann rearrangement which we have just been discussing cannot be used to convert the amide $RCONHCH_3$ into the amine $RNHCH_3$, simply because species analogous to the necessary intermediates $RCONBr^-$ and $RN{=}C{=}O$ (see above) cannot form from the N-methyl-amide.

Similarly, the decarboxylation of β,γ-unsaturated acids takes place

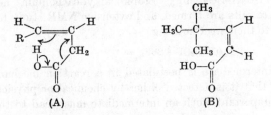

(A)   (B)

readily, and a cyclic transition state (A) has been proposed for this reaction.

The decarboxylation of α,β-unsaturated acids which are also readily decarboxylated is thought to involve prior rearrangement to the β,γ-acids. The acid (B) in which the double bond cannot rearrange to the β,γ-position does not decarboxylate even under the most severe conditions, thus lending support to a mechanism of the type outlined in (A).

In the case of molecular rearrangements we often have to deal with processes in which a fragment is broken off from its position in the reactant molecule and becomes re-attached to a different position in either the same or a different molecule. In considering the mechanism of such reactions it is therefore important to know which of the two possibilities is the correct one, that is, whether the rearrangement is intramolecular or intermolecular. This question can often be answered by carrying out the reaction with a mixture of two similar but non-identical reactants which are known to rearrange at similar rates, and searching the products of the reaction for compounds having fragments of both reactants.

For example, the rearrangement of diazoaminobenzene to *p*-aminoazobenzene undoubtedly occurs by an intermolecular mechan-

$$C_6H_5—N{=}N—NH—C_6H_5 \rightarrow C_6H_5—N{=}N—C_6H_4—NH_2\text{-}p$$
$$\text{Diazoaminobenzene} \qquad\qquad p\text{-Aminoazobenzene}$$

ism because scrambling of the aromatic rings occurs if a mixture of two different diazoaminobenzenes is allowed to rearrange.

Isotopes may also be used to trace the fate of particular atoms as a reaction proceeds. We have already seen how, in favourable cases, mere identification of the products of the reaction normally indicates which bonds are broken and where new bonds are formed. Occasionally, however, more information is required, and this can often be obtained by isotopic labelling.

One of the first applications of isotopes in organic chemistry was the use of oxygen-18 to investigate the mechanism of ester hydrolysis. For example, the hydrolysis of pentyl acetate using water enriched with $H_2^{18}O$ provides unlabelled pentyl alcohol and labelled acetic acid, thus indicating that acyl–oxygen cleavage (a) rather than alkyl–oxygen cleavage (b) has occurred.

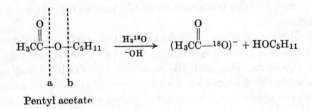

Pentyl acetate

At this point it is worth emphasizing that considerable information as to the intimate details of a mechanism may be gained from stereochemical evidence (pp. 264—272). For instance, hydrolysis of the D-(+)-octan-2-ol acetal of acetaldehyde (A) in dilute aqueous phosphoric acid yields octan-2-ol having the same optical rotation as the original alcohol from which the acetal was synthesized.

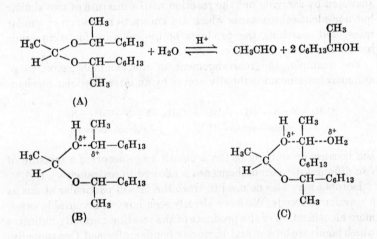

This finding excludes formation of the alkyl carbonium ion [transition state (B)], in which case substantial or complete racemization of the alcohol would be expected, and also a reaction involving nucleophilic attack of solvent on the alcohol [transition state (C)], in which case optical inversion of the alcohol would be expected.

A quite different example of the use of isotopes is seen in the Claisen rearrangement in which allyl phenyl ethers are converted by heat into *o*-allylphenols.

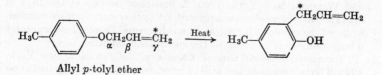

Allyl *p*-tolyl ether

Is the carbon atom that breaks away from the ether linkage the same carbon atom that becomes attached to the benzene ring? The answer is no. If the reaction is carried out using allyl *p*-tolyl ether labelled with ^{14}C at the γ-position, the product is the phenol in which the γ-carbon becomes attached to the benzene ring. The position of labelling can be established by the following degradation sequence.

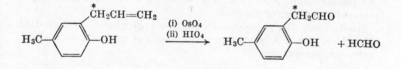

Bibliography

R. P. Bell, *The Proton in Chemistry*, Cornell University Press, Ithaca, New York, 1959. Useful for material on acid and base catalysis.

R. Breslow, *Organic Reaction Mechanisms*, Benjamin, New York, 1966. An excellent introductory treatment of organic reaction mechanisms. This book also includes special-topic chapters which deal with areas of current research interest (e.g. the role of π-complexes in aromatic substitution) in considerable detail.

B. Capon, "Neighbouring Group Participation", in *Quarterly Reviews*, **18**, 45 (1964). A review on neighbouring group participation.

S. L. Friess, E. S. Lewis, and A. Weissberger (Eds.), "Investigation of Rates and Mechanisms of Reactions," in *Technique of Organic Chemistry*, Vol. VII, 2nd edn., Interscience, New York, 1961. Gives a detailed treatment of the methods used to examine reaction mechanisms. Deals with both the experimental details and the mechanistic implications of such techniques as kinetics, detection of intermediates, isotopic labelling, etc.

A. Frost and R. Pearson, *Kinetics and Mechanism*, 2nd edn., Wiley, New York, 1961. An excellent general reference. Particularly good on kinetics and catalysis.

E. S. Gould, *Mechanism and Structure in Organic Chemistry*, Holt, Rinehart and Winston, New York, 1965. A first-class outline of the field of organic reaction mechanisms presented together with a critical examination of the evidence for proposed mechanisms. This text incorporates a large number of exercises at the end of each chapter.

L. P. Hammett, *Physical Organic Chemistry*, McGraw-Hill, New York, 1940. Thirty years old but still of great value. Particularly useful on the study by quantitative methods of the mechanism of reactions and the related problem of the effect of structure and environment on reactivity. (There is a 1970 edition published.)

C. K. Ingold, *Structure and Mechanism in Organic Chemistry*, Cornell University Press, Ithaca, New York, 1953. An account of the early work on reaction mechanisms.

R. O. C. Norman and R. Taylor, *Electrophilic Substitution in Benzenoid Compounds*, Elsevier, New York, 1965. A reappraisal of current theoretical concepts and laboratory practice.

R. Stewart, *The Investigation of Organic Reactions*, Prentice-Hall, Englewood Cliffs, N.J., 1966. A presentation of the topic according to current knowledge of the field.

A. Streitwieser, Jr., *Solvolytic Displacement Reactions*, McGraw-Hill, New York, 1962. A good general discussion of nucleophilic aliphatic substitution reactions.

Problems

1. Ethyl chloride (0.1M) in an acetone solution of potassium iodide (0.1M) is consumed at the rate of 5.44×10^{-7} mole l^{-1} sec^{-1}.

(*a*) If the reaction were to proceed by an S_N2 mechanism, what would the rate of the reaction be at 0.01M concentrations of both reactants?

(*b*) Suppose the rate were proportional to the square of the potassium iodide concentration and the first power of the ethyl chloride; what would the rate be with 0.01M reactants?

(*c*) If one starts with solutions initially 0.1M in both reactants, the rate of formation of ethyl iodide is initially 5.44×10^{-7} mole l^{-1} sec^{-1} but falls as the reaction proceeds and the reactants are used up. Plot the rate of formation of ethyl iodide against the concentration of ethyl chloride as the reaction proceeds (remembering that one molecule of ethyl chloride consumes one molecule of potassium iodide), on the assumption that the rate of reaction is proportional to the first power of the ethyl chloride concentration, and to (i) the zero-th power, (ii) the first power, and (iii) the second power of the potassium iodide concentration.

(*d*) What kind of experimental data would one need to tell whether the rate of the reaction of ethyl chloride with potassium iodide is first-order in each reactant or second-order in ethyl chloride and zero-order in potassium iodide?

2. Suggest mechanisms for the following reactions:

(a)

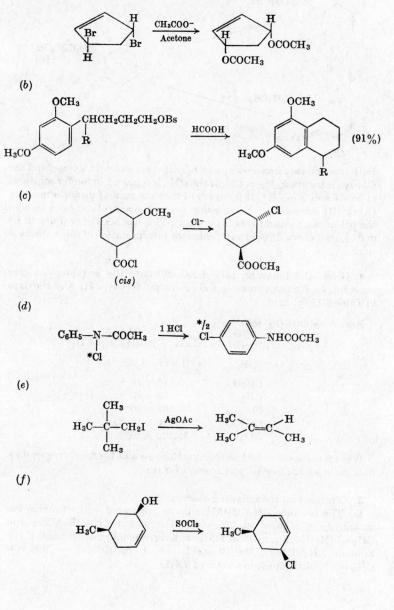

(b)

(c)

(d)

(e)

(f)

3. It has been found that with OH⁻ in 50% aqueous dioxan:

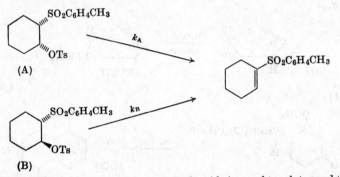

(A)

(B)

Both reactions are first-order in both hydroxide ion and tosylate, and the relative rates are $k_A/k_B = 434$. If the OH⁻ is replaced by buffer solutions of Me_3N and Me_3NH^+, it is observed that the rate of disappearance of (A) or (B) increases with increasing amine concentration, even though the pH of the medium is held constant. This has been shown not to be due to a salt effect. Provide a mechanistic interpretation of these observations.

4. Given the following rate data, calculate the second-order rate constant, k_2, for the unsubstituted compound (i.e. R = H). Use the data in Table 6.11 (p. 321).

Rate constants (k_2) for

$$p\text{-}RC_6H_4CH_2CH_2Br \xrightarrow{\text{NaOEt}} p\text{-}RC_6H_4CH{=}CH_2$$

R	k_2 (l. mole⁻¹ sec⁻¹)
CH_3O	16×10^{-5}
CH_3	23×10^{-5}
Cl	191×10^{-5}
CH_3CO	1720×10^{-5}
NO_2	$75{,}200 \times 10^{-5}$

What is the ρ-value for this reaction? The ρ-value for the corresponding fluoride is +3.12; how do you account for this?

5. Comment on the following observations.

(*a*) The bromide $CH_3CH_2C(Me_2)Br$ on treatment with ethoxide ion in anhydrous ethanol at 70° yields the olefins $CH_3CH{=}CMe_2$ and $CH_2{=}C(Me)Et$ in a ratio of 2.5:1, but corresponding treatment of the bromide $CH_3CH(Et)C(Me_2)Br$ yields the olefins $Et(Me)C{=}CMe_2$ and $CH_2{=}C(Me)CH(Me)Et$ in a ratio of 0.8:1.

(*b*) Apocamphyl chloride (A) is practically inert towards hydroxide ion.

(A)

(*c*) When either of the optical isomers of α-deuterobromobutane is heated with bromide ion it loses its optical rotation just twice as fast as it incorporates radioactive bromine when heated with a large excess of radioactive bromide ion under exactly the same conditions.

(*d*) In the hydrolysis of RX under S_N2 conditions one observes the following rate ratio: Neopentyl/Ethyl = 1/100,000 (i.e. when R = Neopentyl or Ethyl), while for the halide CH_2=CHCH(R)X, the rate ratio is t-Butyl/Methyl = 1/100.

(*e*) The bromination of butan-2-one yields 3-bromobutan-2-one under acid conditions and 1-bromobutan-2-one under basic conditions.

(*f*) Reaction of *trans*-2-acetoxycyclohexyl brosylate with sodium acetate in glacial acetic acid yields mainly the *trans*-diacetate.

(*g*) The *threo*-form of $PhCH(Me)CH(Ph)NMe_3^+I^-$ reacts with sodium ethoxide over fifty times as rapidly as the *erythro*-form.

(*h*) The bromide $C_5H_{11}CH_2CH(Br)CH_3$ reacts with water over thirty times as fast as does the corresponding chloride. The products in both cases are the alcohol $C_5H_{11}CH_2CH(OH)CH_3$ and the olefin $C_5H_{11}CH$=CHCH$_3$, and the ratio of olefin to alcohol in the mixture of products is the same for the bromide as for the chloride, despite the difference in rates.

(*i*) The hydrolysis of benzhydryl chloride, Ph_2CHCl, is more effectively accelerated by addition of Li_2SO_4 than by addition of LiN_3. However, LiN_3 is more effective in diverting the benzhydryl group to benzhydryl azide than is Li_2SO_4 in diverting it to benzhydryl sulphate.

(*j*) In aqueous solution the rate of oxidation by chromic acid of 2-deuteropropan-2-ol, $(CH_3)_2CDOH$, is about one-seventh the rate of oxidation of unlabelled propan-2-ol or of 1,1,1,3,3,3-hexadeutero-propan-2-ol.

(*k*) In the following elimination reaction, the ratio of the reaction rates of the *cis*- and *trans*-compounds is decreased when X = H is replaced by X = NO_2.

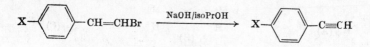

		k	k_{cis}/k_{trans}
X = H	*cis*	3.00×10^{-3}	210,000
	trans	1.4×10^{-8}	
X = NO$_2$	*cis*	3.71	16,000
	trans	2.36×10^{-4}	

(*l*) Terms of the form $k[H_3O^+][AcO^-]$ and $k[AcOH][OH^-]$ do not appear in the general rate expression for bromination of acetone in acetic acid–sodium acetate buffer solution.

(*m*) The relative signs and magnitudes of the *meta* and *para* σ-constants for the substituents Cl and OCH$_3$ are σ_{para}^{OMe} -0.27, σ_{meta}^{OMe} $+0.12$, σ_{para}^{Cl} $+0.23$, and σ_{meta}^{Cl} $+0.37$.

(*n*) The rate constants for the addition of bromine to *para*-substituted styrenes do not correlate well with the Hammett substituent parameters (σ).

6. In the so-called Fries rearrangement, aryl esters such as phenyl acetate are converted into acyl phenols such as *o*-hydroxyacetophenone by the action of Lewis acids. Devise an experiment to show whether this rearrangement is intramolecular or intermolecular.

7. In the Curtius rearrangement, benzoyl azide is converted into nitrogen and phenyl isocyanate. How would you show that the four-membered ring compound (A) is not an intermediate in this reaction?

(A)

8. When the toluenesulphonic ester (A) is dissolved in glacial acetic acid it is converted into the acetate (C). The cyclic ion (B) has been proposed as an intermediate. Devise an experiment that would indicate that such an ion is an active intermediate.

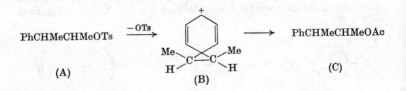

PhCHMeCHMeOTs $\xrightarrow{-OTs}$ → PhCHMeCHMeOAc

(A)   (B)   (C)

9. That the following reaction involves a dehydrobenzene intermediate has been shown by a tracer experiment using ^{14}C (indicated by asterisks). Suggest reaction sequences in which the product is degraded to appropriate compounds, from which the distribution pattern of ^{14}C in the product can be deduced.

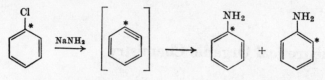

CHAPTER 7

Theoretical Organic Chemistry

We first built up our ideas of organic chemistry in Chapter 1 of Part 1 assuming only the Lewis theory of valency. By Chapter 8 of the first volume we had found it necessary to introduce the theory of 'resonance' and the use of 'canonical forms' to represent molecules which could not be adequately described in terms of the simple electron-pair bonds of the Lewis theory. In the first four chapters of Part 2 we attempted to give a qualitative description of the concepts underlining 'resonance theory', and we also introduced a pictorial 'molecular orbital theory'. The latter did little more than provide a picture; most of the mechanistic chemistry we discussed was still portrayed in terms of resonance theory. It is the purpose of the present chapter to look a little closer at molecular orbital theory. Resonance theory and the use of canonical forms were developed out of the quantum mechanical ideas of 'valence bond theory'. In a somewhat similar way a very useful qualitative theory has been developed out of molecular orbital theory. This theory, called Hückel theory (after its originator), should be regarded as being to molecular orbital theory what resonance theory is to valence bond theory. It is an invaluable guide to the behaviour of conjugated organic molecules, and although we will do some simple calculations these should not be thought of as being any closer to *ab initio* calculations than is the use of canonical forms.

The Linear Combination of Molecular Orbitals

In Chapter 1 of Part 2 we showed that Schrödinger's equation for a particle in one dimension is:

$$\frac{d^2\psi}{dx^2} + \frac{8\pi^2 m}{h^2}(E - V)\psi = 0 \qquad (1)$$

where ψ is the wave function, V the potential energy, and E the energy. In three dimensions the equation becomes (Part 2, p. 6):

$$\nabla^2 \psi + \frac{8\pi^2 m}{h^2}(E - V)\psi = 0 \qquad (2)$$

Let us now rewrite this equation as follows:

$$\left(-\frac{h^2}{8\pi^2 m}\nabla^2 + V\right)\psi = E\psi \qquad (3)$$

The terms inside the brackets are collectively called the *Hamiltonian operator*. We are very familiar with operators although we do not always give them that name. In the equation

$$\frac{d(e^{ax})}{dx} = ae^{ax} \qquad (4)$$

d/dx is an operator, operating on e^{ax}, e^{ax} is known as the *eigen function*, and a as the *eigen value*. The Hamiltonian operator is often abbreviated to $\mathscr{H}$.

$$\left(-\frac{h^2}{8\pi^2 m}\nabla^2 + V\right) = \mathscr{H} \qquad (5)$$

The Schrödinger equation can now be written:

$$\mathscr{H}\psi = E\psi \qquad (6)$$

Notice that if we divide through by ψ we would obtain

$$\psi^{-1}\mathscr{H}\psi = E \qquad (7)$$

i.e. we cannot separate $\mathscr{H}$, the operator, from ψ, the eigen function.

In Chapter 3 of Part 2 we discussed in very qualitative terms the construction of approximate molecular wave functions by simply taking a linear combination of atomic orbitals (LCAO).

$$\psi = \sum_{\nu} c_{\nu}\phi_{\nu} \qquad (8)$$

where ψ represents the molecular wave function while ϕ_{ν} is the atomic wave function of atom ν multiplied by a numerical coefficient c_{ν}. Let us consider a diatomic molecule, with atoms which we shall label as A and B. Then our approximate wave function is:

$$\psi = c_A\phi_A + c_B\phi_B \qquad (9)$$

where c_A and c_B are constants to be determined.

In this equation ψ is the eigen function of a Hamiltonian $\mathbf{H}$.

$$\mathbf{H}\psi = E\psi$$

We need not concern ourselves with the exact nature of $\mathbf{H}$ except to note that it is a one-electron operator since ψ is a one-electron function. Multiplication through by ψ gives:

$$\psi\mathbf{H}\psi = E\psi^2 \tag{10}$$

Integration of (10) over all space gives us:

$$\int \psi\mathbf{H}\psi \, dv = \int E\psi^2 \, dv$$

We have now defined an approximate energy E in terms of our 'one-electron' operator $\mathbf{H}$

$$E = \frac{\int \psi\mathbf{H}\psi \, dv}{\int \psi^2 \, dv} \tag{11}$$

Substituting (9) into (11) gives us:

$$E = \frac{\int (c_A\phi_A + c_B\phi_B)\,\mathbf{H}(c_A\phi_A + c_B\phi_B)\,dv}{\int (c_A\phi_A + c_B\phi_B)^2 \, dv}$$

$$= \frac{\int (c_A\phi_A\mathbf{H}c_A\phi_A + c_A\phi_A\mathbf{H}c_B\phi_B + c_B\phi_B\mathbf{H}c_A\phi_A + c_B\phi_B\mathbf{H}c_B\phi_B)\,dv}{\int (c_A{}^2\phi_A{}^2 + 2c_Ac_B\phi_A\phi_B + c_B{}^2\phi_B{}^2)\,dv} \tag{12}$$

We will simplify this expression by writing:

$$H_{AA} = \int \phi_A\mathbf{H}\phi_A \, dv; \quad H_{BB} = \int \phi_B\mathbf{H}\phi_B \, dv \tag{13a,b}$$

$$H_{AB} = \int \phi_A\mathbf{H}\phi_B \, dv; \quad H_{BA} = \int \phi_B\mathbf{H}\phi_A \, dv \tag{13c,d}$$

$$S_{AA} = \int \phi_A{}^2 \, dv \quad (13e); \quad S_{AB} = \int \phi_A\phi_B \, dv \quad (13f);$$

$$S_{BB} = \int \phi_B{}^2 \, dv \quad (13g)$$

H_{AB} is known as the 'resonance integral', and S_{AB} is known as the 'overlap integral'.

It can be shown that $H_{AB} = H_{BA}$ and so (12) now becomes:

$$E = \frac{c_A{}^2 H_{AA} + 2c_A c_B H_{AB} + c_B{}^2 H_{BB}}{c_A{}^2 S_{AA} + 2c_A c_B S_{AB} + c_B{}^2 S_{BB}} \qquad (14)$$

Now the 'variation theorem' (see *Valence Theory*, J. N. Murrell, S. F. Kettle, and J. M. Tedder, Wiley, London and New York, Chapter VI) states that the energy of an approximate wave function will always be greater than the lowest eigen value of the Hamiltonian. Thus we wish to compute the minimum value of E, i.e. in the general case we require:

$$\frac{\partial E}{\partial c_\nu} = 0 \qquad (15)$$

For our diatomic molecule AB, remembering that

$$\frac{d\left(\dfrac{u}{v}\right)}{dx} = \frac{v\dfrac{du}{dx} - u\dfrac{dv}{dx}}{v^2}$$

we have:

$$\frac{\partial E}{\partial c_A} = \frac{(c_A{}^2 S_{AA} + 2c_A c_B S_{AB} + c_B{}^2 S_{BB})(2c_A H_{AA} + 2c_B H_{AB})}{(c_A{}^2 S_{AA} + 2c_A c_B S_{AB} + c_B{}^2 S_{BB})^2} -$$
$$\frac{(c_A{}^2 H_{AA} + 2c_A c_B H_{AB} + c_B{}^2 H_{BB})(2c_A S_{AA} + 2c_B S_{AB})}{(c_A{}^2 S_{AA} + 2c_A c_B S_{AB} + c_B{}^2 S_{BB})^2} = 0$$

Hence,

$$(c_A{}^2 S_{AA} + 2c_A c_B S_{AB} + c_B{}^2 S_{BB})(2c_A H_{AA} + 2c_B H_{AB}) =$$
$$(c_A{}^2 H_{AA} + 2c_A c_B H_{AB} + c_B{}^2 H_{BB})(2c_A S_{AA} + 2c_B S_{AB})$$

Therefore,

$$c_A H_{AA} + c_B H_{AB} = \frac{c_A{}^2 H_{AA} + 2c_A c_B H_{AB} + c_B{}^2 H_{BB}}{c_A{}^2 S_{AA} + 2c_A c_B S_{AB} + c_B{}^2 S_{BB}} (c_A S_{AA} + c_B S_{AB})$$

The first term on the right-hand side is just E [see (14)]; hence:

$$c_A H_{AA} + c_B H_{AB} = E(c_A S_{AA} + c_B S_{AB})$$

$$\therefore c_A(H_{AA} - S_{AA}E) + c_B(H_{AB} - S_{AB}E) = 0 \qquad (16)$$

We obtain an exactly analogous equation by minimizing E with respect to c_B (i.e. $\partial E/\partial c_B = 0$).

$$c_A(H_{AB} - S_{AB}E) + c_B(H_{BB} - S_{BB}E) = 0 \qquad (17)$$

(16) and (17) are two simultaneous equations, known as the 'secular equations'. Like all simultaneous equations we can express them in determinant form

$$\begin{vmatrix} c_A(H_{AA} - S_{AA}E) & c_B(H_{AB} - S_{AB}E) \\ c_A(H_{AB} - S_{AB}E) & c_B(H_{BB} - S_{BB}E) \end{vmatrix} = 0 \tag{18}$$

Since c_A and c_B are numbers which are not zero, equation (18) can only be satisfied if

$$\begin{vmatrix} H_{AA} - S_{AA}E & H_{AB} - S_{AB}E \\ H_{AB} - S_{AB}E & H_{BB} - S_{BB}E \end{vmatrix} = 0 \tag{19}$$

Equation (19) is known as the 'secular determinant'. Solution of this equation gives us permitted values of E, and substitution of the values into the secular equations gives us the values of the coefficients c_A and c_B.

Returning to the general case, we have an approximate wave function (where μ, ν, π, and ρ are successive atoms)

$$\psi_\mu = \sum_\mu c_\mu \phi_\mu \quad [\phi_\mu \phi_\nu \phi_\pi \phi_\rho \ldots \phi] \tag{8}$$

Substituting (8) into (6) and multiplying through by $c_\nu \phi_\nu$, we have

$$\sum_\mu \sum_\nu c_\mu c_\nu (H_{\mu\nu} - ES_{\mu\nu}) = 0$$

where $H_{\mu\nu} = \int \phi_\mu H \phi_\nu \, dv$ and $S_{\mu\nu} = \int \phi_\mu \phi_\nu \, dv$

Differentiating with respect to, say, c_π, and keeping the other coefficients constant ($H_{\mu\nu}$ and $S_{\mu\nu}$ are constants)

$$2 \sum_\mu c_\mu (H_{\mu\pi} - ES_{\mu\pi}) - \sum_\mu \sum_\nu c_\mu c_\nu S_{\mu\nu} \frac{\partial E}{\partial c_\pi} = 0$$

For an energy minimum,

$$\frac{\partial E}{\partial c_\pi} = 0$$

Therefore,

$$\sum_\mu c_\mu (H_{\mu\pi} - ES_{\mu\pi}) = 0 \tag{20}$$

This is the general secular equation; there are as many equations as there are coefficients.

Hückel Theory; Basic Postulates

Molecular orbital theory is particularly suitable for conjugated molecules because delocalized orbitals are the principal feature of the model. We will consider each carbon atom of a conjugated chain to have an atomic orbital of π-symmetry (the $2p_z$ orbital, say). We shall call these atomic orbitals ϕ_μ, ϕ_ν etc., and the approximate molecular orbital ψ we shall obtain by taking a linear combination of these atomic orbitals:

$$\psi = \sum_\mu c_\mu \phi_\mu \qquad (8)$$

The coefficients and the energies of the molecular orbital are then obtained from the secular equations:

$$\sum_\mu c_\mu (H_{\mu\nu} - E S_{\mu\nu}) = 0 \qquad (20)$$

To reach this secular equation we have already had to make major approximations involved in the LCAO method. In Hückel theory we proceed to make some more major approximations. The Hückel approximations are:

(a) Zero overlap is assumed between atomic orbitals, i.e. $S_{\mu\nu} = 0$ if $\mu \neq \nu$, If we start with normalized atomic orbitals we can take $S_{\mu\mu} = 1$.

(b) $H_{\mu\nu}$, called the 'resonance integral' (see above), is assumed to be the same for all adjacent directly bonded atoms (which we represent as $\mu \to \nu$) and is given the symbol β.

(c) $H_{\mu\mu}$, called the 'Coulomb integral', is assumed to be the same for each carbon $2p\pi$ atomic orbital and is given the symbol α.

(d) $H_{\mu\nu} = 0$ if atom μ is not bonded to atom ν.

The general secular equation of Hückel theory is thus [from eqn. (20)]:

$$c_\mu(\alpha - E) + \sum_{\mu \to \nu} c_\nu \beta = 0$$

Calling the ratio $\dfrac{\alpha - E}{\beta} = m$

$$m c_\mu + \sum_{\mu \to \nu} c_\nu = 0 \qquad (21)$$

The simplest molecule we can consider is ethylene

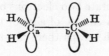

If we label the carbon atoms a and b, the secular equations are

$$(\mu = a) \quad mc_a + c_b = 0$$

$$(\mu = b) \quad c_a + mc_b = 0$$

Hence,

$$\begin{vmatrix} m & 1 \\ 1 & m \end{vmatrix} = m^2 - 1 = 0$$

Therefore, $m = \pm 1$

Hence $E = \alpha + \beta$ or $E = \alpha - \beta$

Substituting $m = -1$ back into the secular equations we have

$$c_a = c_b$$

Applying the normalization condition we have (see Part 2, p. 7)

$$c_a^2 + c_b^2 = 1$$

Therefore, $c_a = \dfrac{1}{\sqrt{2}}$

Now taking the second solution $m = 1$ we find

$$c_a = -c_b \text{ and } c_a = \frac{1}{\sqrt{2}}$$

The two molecular orbitals are thus

$$\psi_1 = \frac{1}{\sqrt{2}}(\phi_a + \phi_b) \quad E_1 = \alpha + \beta$$

$$\psi_2 = \frac{1}{\sqrt{2}}(\phi_a - \phi_b) \quad E_2 = \alpha - \beta$$

The Coulomb integral α represents the energy of an electron in an isolated $2p$ atomic orbital. The resonance integral β is a negative quantity, so that the energy of an electron in ψ_1 is lower than in an isolated p-atomic orbital by this amount, and ψ_1 is therefore *bonding*, whereas the energy of one electron in ψ_2 is greater than in an isolated orbital by the same amount and ψ_2 is therefore *anti-bonding*. In the ground state two electrons occupy ψ_1.

Notice that there is a change of sign of the wave function in the expression for ψ_2, which means there must be a node between the two atoms.

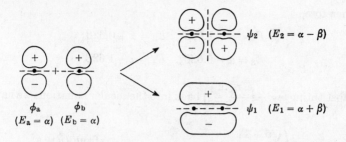

If we treat butadiene in the same way we will just take a linear combination of four atomic orbitals. In Hückel theory we do not distinguish between *cis-* and *trans-*forms.

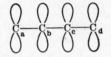

The secular equations for butadiene are

$$(\mu = a) \quad mc_a + c_b \qquad = 0$$
$$(\mu = b) \quad c_a + mc_b + c_c = 0$$
$$(\mu = c) \quad c_b + mc_c + c_d = 0$$
$$(\mu = d) \quad c_c + mc_d \qquad = 0$$

So the secular determinant is

$$\begin{vmatrix} m & 1 & 0 & 0 \\ 1 & m & 1 & 0 \\ 0 & 1 & m & 1 \\ 0 & 0 & 1 & m \end{vmatrix} = 0$$

Therefore,

$$m^4 - 3m^2 + 1 = 0$$

$$\therefore m = \pm \frac{\sqrt{5} \pm 1}{2}$$

Therefore,

$$E_1 = \alpha + 1.62\beta; \quad E_2 = \alpha + 0.62\beta;$$
$$E_3 = \alpha - 0.62\beta; \quad E_4 = \alpha - 1.62\beta$$

Substituting $m = -\left(\dfrac{\sqrt{5}+1}{2}\right)$ back into the secular equations we have

$$c_b = \left(\frac{\sqrt{5}+1}{2}\right)c_a \qquad \text{from } (\mu = a)$$

$$c_c = -c_a + \left(\frac{\sqrt{5}+1}{2}\right)c_b = \left(\frac{\sqrt{5}+1}{2}\right)c_a \quad \text{from } (\mu = b)$$

$$c_c = \left(\frac{\sqrt{5}+1}{2}\right)c_d \quad \text{Therefore, } c_d = c_a \quad \text{from } (\mu = d)$$

From the normalization conditions we have

$$c_a{}^2 + c_b{}^2 + c_c{}^2 + c_d{}^2 = 1$$

hence

$$c_a = \pm\sqrt{2}(6 + 2\sqrt{5})^{1/2} \approx \pm 0.37$$

whence

$$c_b = c_c \approx \pm 0.60$$

so that the lowest-energy bonding orbital is

$$\psi_1 = 0.37\ \phi_a + 0.60\ \phi_b + 0.60\ \phi_c + 0.37\ \phi_d$$

In a similar way we can substitute the other solutions for m and obtain

$$\psi_2 = 0.60\ \phi_a + 0.37\ \phi_b - 0.37\ \phi_c - 0.60\ \phi_d$$
$$\psi_3 = 0.60\ \phi_a - 0.37\ \phi_b - 0.37\ \phi_c + 0.60\ \phi_d$$
$$\psi_4 = 0.37\ \phi_a - 0.60\ \phi_b + 0.60\ \phi_c - 0.37\ \phi_d$$

Notice that the sign of the wave function changes three times in ψ_4, twice in ψ_3, once in ψ_2 and not at all in ψ_1, thus indicating the number of nodes.

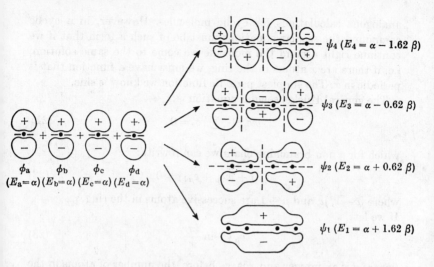

$$\phi_a \qquad \phi_b \qquad \phi_c \qquad \phi_d$$
$$(E_a = \alpha)(E_b = \alpha)(E_c = \alpha)(E_d = \alpha)$$

$\psi_4 \ (E_4 = \alpha - 1.62\ \beta)$

$\psi_3 \ (E_3 = \alpha - 0.62\ \beta)$

$\psi_2 \ (E_2 = \alpha + 0.62\ \beta)$

$\psi_1 \ (E_1 = \alpha + 1.62\ \beta)$

There are four electrons, one from each carbon atom, to put into the four molecular orbitals. They will go in pairs (with opposing spins) into ψ_1 and ψ_2. The total π-electron energy will therefore be $2(\alpha + 1.62\beta) + 2(\alpha + 0.62\beta) = 4\alpha + 4.48\beta$. For two isolated ethylene bonds the energy would be $4\alpha + 4\beta$. The difference, 0.48β, represents the stabilization resulting from the increased electron delocalization in butadiene.

Notice that the energies of the orbitals occur in pairs $\alpha \pm m\beta$, and that the coefficients of the paired orbitals are either the same or the same with opposite sign (See Chapter 1, p. 3). Molecules in which this occurs are called *alternant* hydrocarbons (See Chapter 1, p. 37, for the confusion this term raises). Alternant hydrocarbons turn out to be any polyene, cyclic or linear, provided that in the cyclic compounds the number of atoms in the ring is even. The carbon atoms can then be designated starred or unstarred such that no two atoms of the same set are ever adjacent. In an odd alternant, i.e. a molecule with an odd number of atoms, the more numerous set is taken to be starred.

Cyclic Compounds and the $4n + 2$ Rule

We could continue and carry out analogous calculations for hexa-1,3,5-triene or for octa-1,3,5,7-tetraene etc. Similarly we could make analogous calculations for cyclic molecules. However, in a cyclic compound the secular equations must be of such a form that if we continue right around the molecule we come to the same solution, i.e. if there are x atoms in the ring, we must have a function that is periodic in x. The simplest periodic function we know is sine.

The general Hückel secular equation is:

$$mc_\mu + \sum_{\mu \to \nu} c_\nu = 0 \tag{21}$$

which for a non-branching chain we can rewrite:

$$c_{\mu-1} + mc_\mu + c_{\mu+1} = 0 \tag{22}$$

where $\mu - 1$, μ, and $\mu + 1$ are successive atoms in the ring. If we let:

$$c_\mu = \sin \frac{2\pi\nu}{x} \tag{23}$$

(where ν is an integer and x is, as before, the number of atoms in the ring) we have an expression for c_μ such that $c_\mu = c_{\mu+x}$, i.e. the condition we laid down as essential.

Inserting (23) in (22) we have

$$\sin\left(\frac{2\pi\nu}{x}\right)(\mu - 1) + m\sin\left(\frac{2\pi\nu}{x}\right)\mu + \sin\left(\frac{2\pi\nu}{x}\right)(\mu + 1) = 0$$

Since $\sin A + \sin B = 2 \sin \frac{1}{2}(A + B) \cos \frac{1}{2}(A - B)$ we have

$$2\sin\left(\frac{2\pi\nu}{x}\right)\mu \cos\left(\frac{2\pi\nu}{x}\right) + m\sin\left(\frac{2\pi\nu}{x}\right)\mu = 0$$

Therefore,

$$m = -2\cos\left(\frac{2\pi\nu}{x}\right)$$

and

$$E = \alpha + 2\beta\frac{2\pi\nu}{x} \qquad \nu = 0, 1, \ldots \frac{x}{2}(x \text{ even})$$

$$\text{or} \qquad 0, 1, \ldots \frac{x-1}{2}(x \text{ odd})$$

However, we could equally well let

$$c_\mu = \cos \frac{2\pi\nu}{x} \tag{24}$$

and now inserting (24) in (22) we have:

$$\cos\left(\frac{2\pi\nu}{x}\right)(\mu - 1) + m\cos\left(\frac{2\pi\nu}{x}\right)\mu + \cos\left(\frac{2\pi\nu}{x}\right)(\mu - 1) = 0$$

Since $\cos A + \cos B = 2\cos\frac{1}{2}(A + B)\cos\frac{1}{2}(A - B)$, we have

$$2\cos\left(\frac{2\pi\nu}{x}\right)\mu\cos\left(\frac{2\pi\nu}{x}\right) + m\cos\left(\frac{2\pi\nu}{x}\right)\mu = 0$$

Therefore,

$$m = -2\cos\left(\frac{2\pi\nu}{x}\right) \text{ and hence } E = \alpha + 2\beta\frac{2\pi\nu}{x}.$$

Thus for a cyclic polyene we have two solutions with the same energy for all values of ν, except for $\nu = 0$ and for $\nu = x/2$ (when there is an even number of atoms in the chain). This is because the sine solution is 0 in these circumstances. We call two independent solutions of the wave equation with the same energy 'degenerate'. Thus in cyclic polyenes the orbitals occur in degenerate pairs except for the lowest orbital (if x is odd), or for the lowest and highest orbitals (if x is even).

This result is the celebrated $(4n + 2)$ rule which we expressed in a somewhat different fashion in Chapter 1. If a neutral cyclic molecule C_xH_x has an odd number of atoms, then it must contain an odd number of electrons and hence be a free radical. If the cyclic molecule has an even number of atoms there are two cases to consider, molecules where $x = 2, 6, 10, 14, 18 \ldots (x = 4n + 2)$ in which the π-electrons will go in pairs into bonding orbitals, and molecules where $x = 4, 8, 12, 16 \ldots (x = 4n)$ in which the uppermost pair of π-electrons will according to Hund's rule have to occupy with parallel spins a pair of degenerate orbitals (i.e. giving a triplet state). We discussed all this at some length in Chapter 1 and it needs no further development here.

Charge Density

In our preliminary discussion of the wave equation in Part 2, Chapter 1, we interpreted ψ^2 as the 'amount of electron charge' per unit volume. Electron densities play a major role in our discussions of organic reactions. An electron in a molecular orbital ψ_r has a density distribution ψ_r^2.

$$\psi_r = \sum_\mu c_{r\mu}\phi_\mu$$

Now in Hückel theory we neglect overlap density terms, e.g. terms such as $\phi_\mu \phi_\nu$, hence our density distribution becomes $\sum_\mu c_{r\mu}^2 \phi_\mu{}^2$. It follows that $c_{r\mu}^2$ is a measure of the charge density at atom μ contributed by an electron in molecular orbital ψ_r. To get the total π-electron density at atom μ, we sum over all the occupied orbitals (remembering there are two electrons in each). Thus we can define an atom charge density q_μ as:

$$q_\mu = 2 \sum_{r.occ.} c_{r\mu}^2$$

For ethylene (p. 362) we have $q_a = 2c_{1a}^2 = 2\left(\dfrac{1}{\sqrt{2}}\right)^2 = 1$

For butadiene (p. 364) we have

$$q_a = q_d = 2\,c_{1a}^2 + 2\,c_{2a}^2 = 2(0.37)^2 + 2(0.60)^2 = 1$$

and

$$q_b = q_c = 2\,c_{1b}^2 + 2\,c_{2b}^2 = 2(0.60)^2 + 2(0.37)^2 = 1$$

Thus in both ethylene and butadiene we find that the *π-electron density is unity at each atom*. It can be shown that this result is true for all alternants, but it is *not* true for non-alternants. Thus naphthalene and biphenyl are non-polar molecules, but azulene has an appreciable dipole. The charge density distribution of the ground state of an alternant hydrocarbon cannot therefore be important in determining the site of attack by another species.

Simple Lewis theory first broke down when we were considering the acetate anion. We turned instead to resonance theory; let us now re-examine this problem but instead of looking at the acetate anion itself we will examine the allyl anion (this is because our assumption that $H_{\mu\mu} = \alpha$ is the same for all atoms in the chain cannot apply if one of the atoms is a heteroatom, i.e. oxygen). The Hückel secular equations for the allyl system are

$$mc_a + c_b \qquad\quad = 0$$
$$c_a + mc_b + c_c = 0$$
$$c_b + mc_c \qquad = 0$$

The secular determinant is:

$$\begin{vmatrix} m & 1 & 0 \\ 1 & m & 1 \\ 0 & 1 & m \end{vmatrix} = 0$$

Therefore, $m^3 - 2m = 0$

$m(m^2 - 2) = 0$

$m = -\sqrt{2}, 0, +\sqrt{2}$.

The Hückel orbitals are thus:

$$\psi_1 = \tfrac{1}{2}\phi_a + \sqrt{\tfrac{1}{2}}\phi_b + \tfrac{1}{2}\phi_c, \quad E_1 = \alpha + \sqrt{2}\beta$$

$$\psi_2 = \sqrt{\tfrac{1}{2}}\phi_a - \sqrt{\tfrac{1}{2}}\phi_c, \quad E_2 = \alpha$$

$$\psi_3 = \tfrac{1}{2}\phi_a - \sqrt{\tfrac{1}{2}}\phi_b + \tfrac{1}{2}\phi_c, \quad E_3 = \alpha - \sqrt{2}\beta$$

In the allyl cation we have two π-electrons both of which will go into ψ_1. Hence $q_a = q_c = 2c_{1a}^2 = \tfrac{1}{2}$; $q_b = 2c_{1b}^2 = 1$. In the anion the two further electrons must be placed in ψ_2; hence $q_a = q_c = 2c_{1a}^2 + 2c_{2a}^2 = 1\tfrac{1}{2}$; $q_b = 2c_{1b}^2 + 2c_{2b}^2 = 1$, i.e.

the cation has charge density $\overset{\frac{1}{2}+}{C_a}—C_b—\overset{\frac{1}{2}+}{C_c}$

and the anion $\overset{-\frac{1}{2}}{C_a}—C_b—\overset{-\frac{1}{2}}{C_c}$

This is exactly the same answer as we would get from resonance theory. We could have reached this result without going through the complete Hückel calculation. We described above how in an 'alternant' hydrocarbon the energy levels of the Hückel orbitals occurred in pairs with energy $E = (\alpha + m\beta)$ (*bonding*) and $E = (\alpha - m\beta)$ (*antibonding*). It follows that if the molecule has an odd number of atoms in the conjugated chain there must be a 'non-bonding orbital' with energy $E = \alpha$ (i.e. for which $m = 0$).

If we now return to the general Hückel secular equation:

$$mc_\mu + \sum_{\nu \to \mu} c_\nu = 0$$

we see that if $m = 0$ the sum of the coefficients around any one atom must be zero. Now in an 'alternant' molecule starred atoms are directly bonded only to unstarred atoms, hence the coefficients of either the starred or the unstarred set must be zero. Examination shows that this must be the smaller (the 'unstarred') set. We can now obtain the coefficients of the non-bonding orbital of any odd alternant very easily.

Let us look at allyl on which we have already done the full Hückel
calculation, and consider the coefficients of the non-bonding orbital,
$c_{0\mu}$.

$$\overset{*}{C_a}—C_b—\overset{*}{C_c}$$

We see at once that $c_{ob} = 0$. Let us put $c_{oa} = a$, then summing around
atom 2 we see that $c_{oc} = -a$. If we normalize this we get $a^2 + (-a^2) =
2a^2 = 1$, hence $a = \sqrt{\frac{1}{2}}$, giving $c_{oa} = \sqrt{\frac{1}{2}}$ exactly the same answer as
we obtained before. We obtain the π-electron charges of the anion
by adding one electron to the non-bonding orbital which therefore
converts the neutral molecule into

$$\overset{-\frac{1}{2}}{C_a}—C_b—\overset{-\frac{1}{2}}{C_c}$$

Let us now consider a more complicated example. In Chapter 14,
p. 142, of Part 1 we depicted the anion of phenol as follows (see also
p. 64 of Part 2).

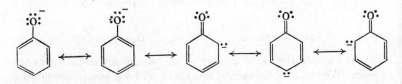

We can treat this ion in exactly the same way as we dealt with the
acetate ion. We begin by taking the benzyl anion as a model to avoid
the problems associated with the extra electronegativity of the oxygen
atom just as in the case of acetate we examined allyl. The next step
is to star the most numerous set of atoms as before:

Benzyl, an odd alternant

Now let us assume the coefficient of the non-bonding orbital for the
para-carbon atom is a ($c_{04} = a$), then the coefficients of the other
starred atoms must be as follows:

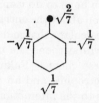

Normalizing we get

$$7a^2 = 1$$

$$a = \frac{1}{\sqrt{7}}$$

which gives us coefficients for the non-bonding orbital as follows

Coefficients of the non-bonding orbital of benzyl

To obtain the charge densities of the anion we just add one electron to the non-bonding orbital and so the charge density is given by $q_\mu = c_{0\mu}^2$ (where subscript o indicates the non-bonding orbital).

Charge densities in the benzyl anion
(a model for the phenoxide ion)

For a diazonium salt we take one electron away from the system, i.e. we regard the benzyl cation as our model and hence the charge distribution is the same though of the opposite sign.

Charge densities in the benzyl cation

Reactivity Indices and Localization Theory

Valency theories can tell us nothing directly about the reactivities of molecules. Really to calculate reaction rates we need to calculate the free energy change in going from reactants to the transition state, and the most that we can hope to do using valency theories alone is to obtain some information about the enthalpy changes. The simplest thing we could do is to calculate charge densities. But we have seen that for all alternant hydrocarbons the charge density is unity at every atom. The Japanese chemist Fukui has suggested that in a reaction between an electrophile and an olefin the greatest part of the energy change is due to the interaction between the lowest anti-bonding or non-bonding orbital of the electrophile with the highest bonding orbital of the hydrocarbon. These are called the 'frontier orbitals'. He then suggests that only the electron density in the highest bonding orbital, 'the frontier orbital', need be considered. For example the π-electron densities at carbon atoms 1 and 2 in butadiene due to the two electrons in ψ_2 (the frontier orbital) are approximately 0.73 and 0.28 respectively (see p. 364). Hence we predict that an electrophile will attack carbon atom 1 (or carbon atom 4) preferentially. In nucleophilic attack we would consider interaction between the highest occupied orbital of the nucleophile and the lowest unoccupied orbital of the hydrocarbon. For butadiene the numerical values of the Hückel coefficients of ψ_3 are identical with those of ψ_2 so that π-electron densities in ψ_3 (involving $c_{3,\mu}^2$) must be the same. Thus nucleophilic addition, if it did occur, should also be at the terminal atoms. Frontier electron theory is very hard to justify on theoretical grounds but it has the merits of simplicity.

The most used and probably the soundest based reactivity indices are localization energies. In localization theory one attempts to assess the loss of π-electron energy in forming the initial adduct in an addition to an olefin or an aromatic compound. For example in an addition-with-elimination reaction on benzene the first step is the formation of a Wheland intermediate, in which a pair of electrons

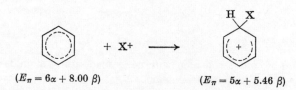

$(E_\pi = 6\alpha + 8.00\ \beta)$ $(E_\pi = 5\alpha + 5.46\ \beta)$

have been localized (at carbon atom μ say) to form a bond with X. The total π-electron energy for benzene according to Hückel theory is $6\alpha + 8.00\beta$ while the total π-electron energy for the pentadienyl cation (i.e. the remaining delocalized system) is $5\alpha + 5.46\beta$. Thus the *localization energy* $L_\mu{}^+$ for benzene is 2.54β.

For alternant hydrocarbons there is an easy way of obtaining approximate localization energies. Dewar has pointed out that any even alternant hydrocarbon RS can be broken down into two odd alternant fragments, R and S. Thus benzene could be regarded as two allyl fragments joined together. Dewar showed that the total Hückel energy of RS is related to the sum of the energies of R and S by the following approximate relation:

$$E_{\mathrm{RS}} \approx E_{\mathrm{R}} + E_{\mathrm{S}} + 2\beta \left(\sum_{\mu \to \nu} c^{\mathrm{R}}_{0\mu} c^{\mathrm{S}}_{0\nu} \right)$$

where $c^{\mathrm{R}}_{0\mu}$ and $c^{\mathrm{S}}_{0\nu}$ are the coefficients of the non-bonding orbitals of R and S respectively for atoms μ (of R) and ν (of S) which are joined. The summation is over all $\mu \to \nu$ bonds. We have seen that the coefficients of the non-bonding orbitals of allyl are $\pm 1/\sqrt{2}$, and we can also calculate the Hückel energy of the radical from the results on p. 369 to be $3\alpha + 2.82\beta$. Thus if we construct benzene from two allyl fragments:

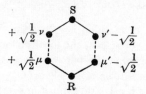

We have

$$E_{RS} \approx (3\alpha + 2.82\beta) + (3\alpha + 2.82\beta) + 2\beta[(\sqrt{\tfrac{1}{2}} \times \sqrt{\tfrac{1}{2}}) + (-\sqrt{\tfrac{1}{2}} \times -\sqrt{\tfrac{1}{2}})]$$

$$\qquad\quad E_R \qquad + \qquad E_S \qquad + \qquad c_{0\mu}^{R} \qquad + \qquad c_{0\nu}^{S}$$

$$\approx 6\alpha + 7.63\beta$$

We could of course consider benzene as made up of a single carbon atom and a pentadienyl fragment:

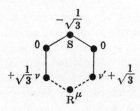

We now have

$$E_{RS} \approx \alpha + 5\alpha + 5.46\beta + 2\beta\left(1 \times \frac{1}{\sqrt{3}} + 1 \times \frac{1}{\sqrt{3}}\right)$$

$$\approx 6\alpha + 7.77\beta$$

These two approximate calculations must be compared with $E = 6\alpha + 8.0\beta$ for the full Hückel calculation. However, it is not for the calculation of total Hückel energies that this approximation is important but for its use in calculating localization energies. In the last example in which one fragment is a single atom we have a model analogous to the Wheland intermediate in addition-with-elimination reactions. Now the coefficients of the single atom $c_{0\mu}^{R} = 1$ and the two atoms of fragment S, namely ν and ν' which are joined to μ have coefficients $c_{0\nu}$ and $c_{0\nu'}$; hence we can write:

$$E_{RS} - E_R - E_S \approx 2\beta(c_{0\nu} + c_{0\nu'})$$

Thus the sum $2(c_{0\nu} + c_{0\nu'})$ which is called the 'Dewar Number' N_{μ}, when multiplied by β represents the localization energy. The smaller N_{μ} the smaller the interaction between R and S and the easier it will be to isolate atom μ from the π-electron system. Let us compare electrophilic attack at the 1- and 2-positions in naphthalene:

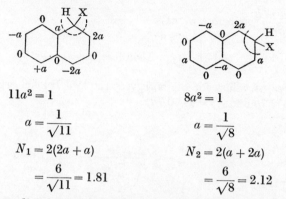

$$11a^2 = 1 \qquad\qquad 8a^2 = 1$$

$$a = \frac{1}{\sqrt{11}} \qquad\qquad a = \frac{1}{\sqrt{8}}$$

$$N_1 = 2(2a + a) \qquad\qquad N_2 = 2(a + 2a)$$

$$= \frac{6}{\sqrt{11}} = 1.81 \qquad\qquad = \frac{6}{\sqrt{8}} = 2.12$$

Thus we predict that in naphthalene attack will take place preferentially at the 1-position (i.e. the position with the lowest Dewar number N). This prediction is, of course, in accord with experimental observation.

Heteroatoms

So far we have only discussed molecules consisting entirely of carbon and hydrogen atoms (we have, in fact, ignored the latter). The introduction of oxygen or nitrogen atoms into an unsaturated system will require new values for both the coulomb and resonance integrals. Changes in α and β are usually expressed in terms of the values used for carbon atoms in benzene, which we can take as α_C and β_{C-C}. For heteroatom X we have

$$\alpha_X = \alpha_C + h_X \beta_{C-C}$$
$$\beta_{C-X} = k_X \beta_{C-C}$$

It is extremely difficult to decide on the best way of determining suitable values of h_X and k_X. In practice it is probably best to vary their values to obtain the best fit with experiment. In general the more electronegative X the larger h_X and k_X. To illustrate the kind of result you obtain, we will compare formaldehyde with ethylene. We will take $h_{=O} = 2$ and $k_{C=O} = \sqrt{2}$ (notice that the values of h_X and k_X will be different for oxygen in a double bond from those in a single bond). The secular equations for formaldehyde are thus:

$$c_1(\alpha_C - E) + c_2 k_{C=O}\beta_{C-C} = 0$$
$$c_1 k_{C=O}\beta_{C-C} + c_2(\alpha_C + h_{=O}\beta_{C-C} - E) = 0$$

If $m = \dfrac{\alpha_C - E}{\beta_{C-C}}$ as before we have

$$\begin{vmatrix} m & \sqrt{2} \\ \sqrt{2} & m+2 \end{vmatrix} = 0 \qquad \text{Therefore, } m = 0.7; -2.7$$

This gives us values for the energy of the two Hückel orbitals of $E_1 = \alpha + 2.7\beta$ and $E_2 = \alpha - 0.7\beta$.

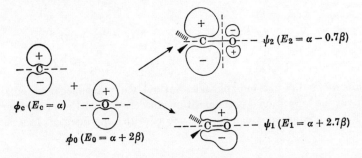

Notice that the bonding molecular orbital is close in energy to the oxygen atomic orbital, and the antibonding molecular orbital is close in energy to the carbon atomic orbital. Compare these results with those we obtained for ethylene (p. 362).

Hückel calculations involving heteroatoms are never as satisfactory for quantitative prediction as those involving purely carbon atom systems, but they can sometimes give useful insight into a molecule's behaviour. We can see, for example, that the above treatment of formaldehyde is consistent with all that we know about the chemistry of the carbonyl group.

Electronic Spectra and Excited States

In Part 2, Chapter 1, we discussed absorption of light by an atom. We noted that an atom would only absorb light of certain specific wavelengths and we associated the absorption of a packet or 'quantum' of energy with the promotion of an electron from one orbital to another of higher energy. The difference in energy between the two states is given by the relation

$$E_2 - E_1 = h\nu$$

where h is Planck's constant and ν is the frequency of the energy absorbed. In Chapter 25 of the same volume we considered the absorption of visible and ultraviolet light by organic molecules. We concluded that in a molecule, absorption of ultraviolet light was associated with the promotion of an electron from one *molecular orbital* to another. We considered among other examples the electronic spectra of some linear polyenes, and we attributed the observed spectrum to the promotion of an electron in the highest bonding π-orbital to the lowest antibonding π-orbital (so-called $\pi-\pi^*$ transitions). We can now, using the Hückel method discussed above, calculate the energy differences between the highest bonding and lowest antibonding π-orbitals in terms of the resonance integral β.

Energy difference between highest bonding and
lowest antibonding orbitals in linear polyenes

	E_1	E_2	$E_2 - E_1$
Ethylene	$\alpha - \beta$	$\alpha + \beta$	2β
	E_2	E_3	
Butadiene	$\alpha + 0.62\beta$	$\alpha - 0.62\beta$	1.24β
	E_3	E_4	
Hexatriene	$\alpha + 0.45\beta$	$\alpha - 0.45\beta$	0.90β
	E_4	E_5	
Octatetraene	$\alpha + 0.35\beta$	$\alpha - 0.35\beta$	0.70β

If we now plot the reciprocal of the observed wavelength of the maxima of the ultraviolet spectra of the linear polyenes against the calculated energy in terms of β, then, remembering that $E_2 - E_1 = \boldsymbol{h\nu}$, we should obtain a straight line whose slope corresponds to $\boldsymbol{h\beta}$.

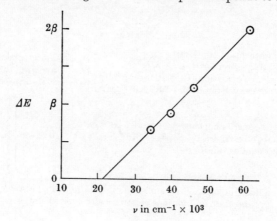

ν in cm^{-1} × 10^3

This gives us a value of $\beta = -248$ kJ mole^{-1} (60.3 kcal mole^{-1}). We could make similar plots for the ultraviolet spectra of aromatic hydrocarbons, or for unsaturated aldehydes. In both cases we would obtain good straight lines but the slopes would be different (ca. -260.5 kJ mole^{-1}, i.e. 62.3 kcal mole^{-1}, for aromatic hydrocarbons and ca. -296 kJ mole^{-1}, i.e. -71 kcal mole^{-1}, for aldehydes).

The fact that we obtain good linear relations between observed ultraviolet absorption maxima and calculated Hückel energy levels *within one series* tells us something about Hückel theory. It shows that qualitatively Hückel theory gives us the correct picture for a given series of analogous compounds. At the same time it shows that quantitatively Hückel theory is not reliable; in fact, if we take different types of experiment we find β can vary by a factor of nearly 5. The reason for this is that in Hückel theory we ignore electron-interaction terms. For any particular experimental quantity involving similar molecules the effect of ignoring these terms is usually similar. Once

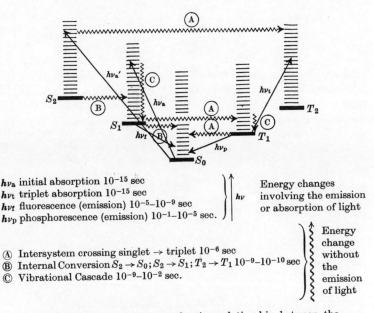

$h\nu_a$ initial absorption 10^{-15} sec
$h\nu_t$ triplet absorption 10^{-15} sec
$h\nu_f$ fluorescence (emission) 10^{-5}–10^{-9} sec
$h\nu_p$ phosphorescence (emission) 10^{-1}–10^{-5} sec.

$\left.\right\}$ $h\nu$ Energy changes involving the emission or absorption of light

Ⓐ Intersystem crossing singlet → triplet 10^{-6} sec
Ⓑ Internal Conversion $S_2 \rightarrow S_0$; $S_2 \rightarrow S_1$; $T_2 \rightarrow T_1$ 10^{-9}–10^{-10} sec
Ⓒ Vibrational Cascade 10^{-9}–10^{-2} sec.

$\left.\right\}$ Energy change without the emission of light

Modified Jablonski diagram showing relationship between the ground state and various electronically excited states we change either the type of compound or the experiment, the electron-interaction terms cannot be neglected and Hückel theory will only give the very approximate results.

When an electron is excited from one molecular orbital to another the spin of the electron remains unchanged. In butadiene the lowest absorption band corresponds to the excitation of one of the two electrons in ψ_2 to ψ_3. This is an excited singlet state designated S_1. Each electronic state has its series of quantized vibrational levels and each vibrational level is further split into quantized rotational levels (not shown).

The initial absorption of light $h\nu$ may be followed by the return of the electron to its original orbital with the emission of light ($h\nu_f$—fluorescence). Owing to what is known as spin–orbit interaction the spin of the electron can in certain circumstances invert, and the resultant triplet state is often of lower energy than the first excited singlet state.

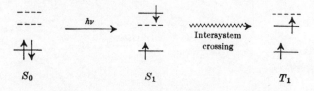

Simplified diagram of singlet and triplet excited states

What is particularly important from a chemical point of view is that the lifetime of the triplet is long in comparison with the excited singlet, and chances of reaction are consequently very much greater. The triplet may also have a different preferred conformation; thus in ethylene the triplet state has a preferred conformation in which the terminal bonds are at right angles to one another.

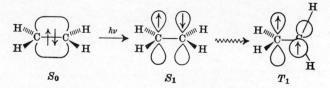

Not surprisingly *cis–trans* isomerization is often promoted by light.

The electron densities as well as the molecular geometry of excited molecules will differ from those of the ground state molecules. At the present time much work is being devoted to the study of photochemical reactions. So far there is a fairly wide gulf between physical studies on the nature of the primary processes and chemical studies on the nature of the ultimate products. The chemistry of photo-excited molecules (photochemistry) will not be discussed in the present volume; full discussion of these topics will be found in a number of excellent monographs.

Pericyclic Reactions and the Conservation of Orbital Symmetry (The Woodward–Hoffmann Rules)

Let us consider an electrocyclic ring opening reaction, for example, the thermal isomerization of dimethyl *cis*-cyclobutene-1,2-dicarboxyl-ate to yield dimethyl *trans,cis*-butadiene-1,4-dicarboxylate.

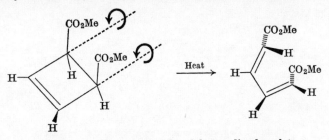

Conrotatory ring-opening of cyclobutenedicarboxylate

This reaction is stereospecific; notice that the carboxylate groups have rotated in the same direction, such a process is called 'conrotatory'. If we now look at the *frontier orbitals* involved we have the following picture:

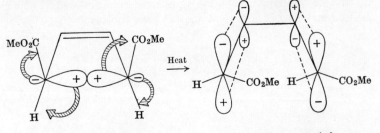

σ Orbital to be broken

ψ_2 the highest occupied π orbital of butadiene

Notice that a bonding interaction between these frontier orbitals is maintained at all stages of the reaction. Such a process is said to be symmetry-allowed, i.e. orbital symmetry is conserved during the whole process (by orbital symmetry we refer to the relative signs of the wave function).

Let us look at another example; *trans,cis,trans*-1,6-dimethylhexa-1,3,5-triene cyclizes on heating to *cis*-1,2-dimethylcyclohexa-3,5-diene, but on irradiation the same triene cyclizes to yield the *trans*-dimethylcyclohexadiene:

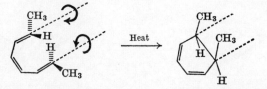

Disrotatory—thermal cyclization of dimethylhexatriene

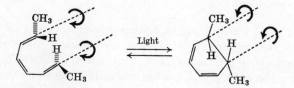

Conrotatory—photochemical cyclization of dimethylhexatriene

Both of these reactions are stereospecific. Notice that in the photochemical cyclization of the hexatriene (which is reversible) the terminal groups rotate the same way, i.e. the reaction is conrotatory like the thermal ring opening of the cyclobutene. In the thermal ring closure of the hexatriene, however, the terminal groups rotate in the opposite direction and we call the reaction 'disrotatory'.

We can depict the frontier orbitals of those two processes as shown at top of page 382.

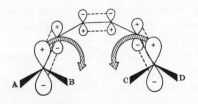

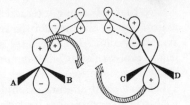

ψ_3 The highest occupied π-orbital of the ground state of hexatriene

ψ_4 The highest occupied π-orbital of the 1st excited state of hexatriene

Thermal

Photochemical

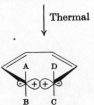

σ Bond of cyclohexadiene formed in cyclization

Disrotatory

σ Bond of cyclohexadiene formed in cyclization

Conrotatory

From our knowledge of the pairing properties of Hückel orbitals we know that the number of nodes in the highest bonding orbital of a linear polyene will be $(n/2) - 1$ where n is the number of carbon atoms in the chain (for a neutral molecule n must of necessity be even). If $n/2$ is odd ($n = 6, 10, \ldots$) the symmetry of the highest bonding orbitals is such that a disrotatory process will lead to a bonding σ-bond between the terminal carbon atoms on ring closure. If $n/2$ is even ($n = 4, 8, \ldots$) the symmetry of the highest bonding orbital is such that a conrotatory process is necessary to produce a bonding σ-bond between the terminal carbon atoms on ring closure.

It is possible to extend this argument and consider all the orbitals, not just the frontier orbitals. This involves the use of symmetry operations described in Chapter 23 of Part 2. We will not go through these arguments in the present volume as they are fully developed elsewhere (J. N. Murrell, S. F. Kettle, and J. M. Tedder, *Valence Theory*, 2nd Edn., Wiley, London and New York, 1970).

It is very important to appreciate that these arguments assume that the reactions occur as a concerted process; they will not apply to a reaction that occurs in discrete steps.

The electrocyclic reactions we have been discussing are closely related to the cycloaddition reactions of the Diels–Alder type. The striking feature of these reactions is that some go thermally with ease, and others which will not go thermally will go under the influence of light, e.g.

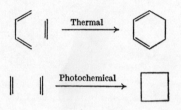

These reactions occur in a stereospecific manner and involve an activated complex in which the interacting molecules are in parallel planes.

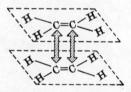

Applying our ideas for the conservation of orbital symmetry, we find that such a process will be allowed thermally (i.e. ground state interaction) if half the total number of atoms (q) of the two fragments containing m atoms and n atoms respectively is odd (i.e. $m + n = q$; cycloaddition thermally allowed if $q/2$ is odd). The situation is reversed for a photochemical process (i.e. $m + n = q$; cycloaddition thermally forbidden but photochemically allowed if $q/2$ is even).

We have said that these cycloaddition reactions involve the two molecules approaching each other in parallel planes; thus, for ethylene we would depict the process as follows:

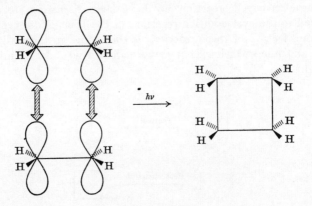

We could however, visualize the interaction between two ethylenes with their molecular axes at right angles to one another.

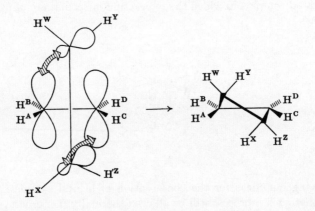

When the ethylenes approach in parallel planes the orbital lobes come from the same side of the molecular plane in each molecule, but when the ethylenes approach at right angles the interacting lobes come from opposite sides of the molecular plane for one of the ethylenes. We can depict these two situations as follows:

(a) If we are making or breaking bonds on the same side of a

π-orbital nodal plane we call the process 'suprafacial' and designate it S (notice that S is similar to '*syn*' and 'synclinal' but *not* the same).

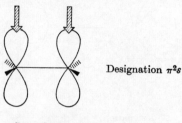

Designation $\pi^2 s$

Suprafacial

(*b*) If we are making or breaking bonds on opposite sides of a π-orbital nodal plane we call the process 'antarafacial' and designate it A (notice that A is similar to '*anti*' and 'anticlinal' but *not* the same).

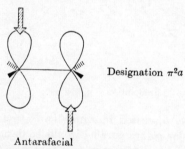

Designation $\pi^2 a$

Antarafacial

We can now visualize these types of cycloaddition between two conjugated polyolefins with m atoms and n atoms in their carbon chains.

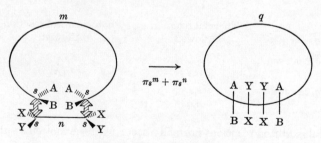

designation $= \pi_s{}^m + \pi_s{}^n$

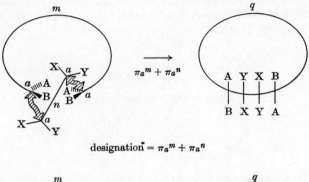

designation $= \pi_a{}^m + \pi_a{}^n$

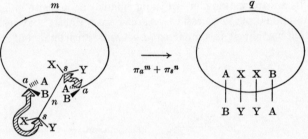

designation $= \pi_a{}^m + \pi_s{}^n$

The reaction we discussed in which two ethylenes approached with their molecular planes parallel would be designated $\pi_s{}^2 + \pi_s{}^2$, and the reaction in which two ethylenes approached with molecular planes at right angles would be designated $\pi_s{}^2 + \pi_a{}^2$. The rules we obtain for cycloaddition reactions involving two polyolefins (one with m atoms and one with n) when orbital symmetry is considered are:

$q = m + n$	Thermal reaction Allowed	Photochemical reaction Allowed
$q/2$ is even	s,a	s,s
	a,s	a,a
$q/2$ is odd	s,s	s,a
	a,a	a,s

Cycloaddition reactions normally occur with the molecular planes of the interacting molecules parallel, probably because of steric hindrance to the right-angle approach. Thus ethylene dimerization can be represented as follows:

$$\pi_s{}^2 + \pi_s{}^2$$
$$q = 4 \quad q/2 \text{ is even}$$

Therefore the symmetry-allowed process is photochemical.

And a Diels–Alder reaction similarly:

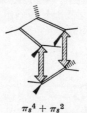

$$\pi_s{}^4 + \pi_s{}^2$$
$$q = 6 \quad q/2 \text{ is odd}$$

Therefore the symmetry-allowed process is thermal.

So far we have only considered π-orbitals but σ-orbitals can be treated in the same way. The new bond can be formed with retention or inversion at the tetravalent carbon atoms. When the atoms at both ends of the σ-bond both retain their configuration or both invert their configuration, we designate the process suprafacial (σ_s).

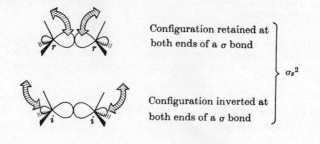

Configuration retained at both ends of a σ bond

Configuration inverted at both ends of a σ bond

$\sigma_s{}^2$

When an atom at one end of the σ-bond retains configuration but
that at the other end inverts, we designate the process antarafacial (σ_a).

Configuration retained at
one end of a σ bond but   } $\sigma_a{}^2$
inverted at the other end

To understand the implications of this argument we shall go back
and reconsider the electrocyclic ring opening reactions we discussed
initially. For cyclobutene the symmetry-allowed process is conrotatory and we can depict this as either suprafacial at the π-orbital
and antarafacial at the σ-orbital, *or* antarafacial at the π-orbital
and suprafacial at the σ-orbital.

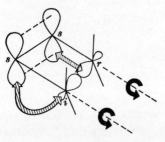

$\pi_s{}^2 + \sigma_a{}^2$

$q = 4$. Therefore $q/2$ is even.

Therefore *thermal s,a* process
is symmetry-allowed.

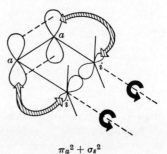

$\pi_a{}^2 + \sigma_s{}^2$

$q = 4$. Therefore $q/2$ is even.

Therefore *thermal a,s* process
is symmetry-allowed.

Two ways of depicting a conrotatory ring opening of a cyclobutene, in which the orbital symmetry of both the π-orbitals
and σ-orbitals being broken are considered simultaneously.

The photochemical rearrangement of a cyclohexadienone can be
pictured in a similar way.

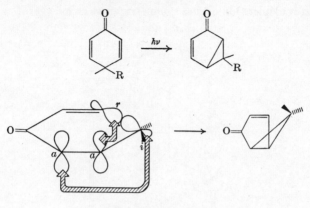

$$\sigma_a{}^2 + \pi_a{}^2 \quad q/2 \text{ is even}$$

Therefore an *a,a photochemical* process is symmetry-allowed (notice that an *s,s* process is stereochemically impossible).

These ideas can be extended to the treatment of conjugated ions with an odd number of atoms (in which a carbonium ion has a vacant *p*-atomic orbital designated ω^0 and a carbanion has a completely filled orbital designated ω^2; see problems, p. 391). They can also be applied to sigmatropic rearrangements. A sigmatropic rearrangement is the name given to 1,3- or 1,5- or 1,7-carbon to carbon (or hydrogen to carbon) shifts.

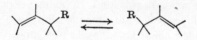

Sigmatropic shift

An example of such a thermal rearrangement is depicted below

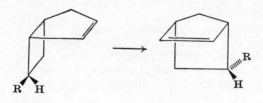

and can be depicted in orbital symmetry terms as follows:

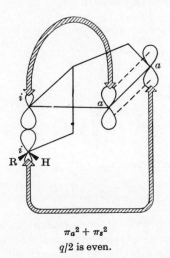

$$\pi_a{}^2 + \pi_s{}^2$$
$q/2$ is even.

Thus *thermal a,s* process is symmetry-allowed

It will be seen that these ideas involving the relative symmetry properties of interacting orbitals can be applied to a very wide range of reactions. The argument often involves 'frontier orbitals' and the theoretical justification is difficult to find. *The treatment can only apply to concerted processes*; reactions which go in steps cannot be governed by arguments of this kind. In photochemical reactions the multiplicity of the excited state need not alter the orbital symmetry arguments provided that the subsequent changes occur in a concerted fashion.

This is a field of organic chemistry undergoing very rapid growth. The present discussion deals with the basic concepts; applications of these ideas must be sought in current literature.

Bibliography

This chapter has been deliberately written to correlate with *Valence Theory*, by J. N. Murrell, S. F. A. Kettle, and J. M. Tedder, Wiley, London and New York, 2nd. Edn., 1970.

The best qualitative description of valency theories remains *Valence*, by C. A. Coulson, Oxford University Press, 2nd. Edn., 1962.

The fullest account of molecular orbital theory in relation to organic chemistry is *Molecular Orbital Theory for Organic Chemists*, by A. Streitwieser, Wiley, 1961.

An excellent introduction to valency theory can be found in Chapter 1, "Wave Mechanics and the Alkene Bond", by C. A. Coulson and E. T. Stewart, in *The Chemistry of the Alkenes* (Ed. S. Patai), Interscience, London and New York, 1964.

Since this chapter was written an excellent account of pericyclic reactions and orbital symmetry has been published by the originators of this approach [R. B. Woodward and R. Hoffmann, *Angew. Chem.*, 8, 781—853 (1969)].

Problems

1. Obtain the Hückel orbitals for cyclo-C_3H_3, and compare the π-electron energy of the cyclopropenylium ion with that of the allyl cation.

2. Calculate the charge densities in the anions of 1- and 2-naphthol (taking the appropriate $C_{10}H_7CH_2$ ions as your model).

3. Determine the Dewar numbers for the available positions in phenanthrene, and so predict the relative rates of electrophilic attack at the various sites.

4. Under what conditions would you expect butadiene to dimerize, and what would you predict as the product?

5. Discuss the possible modes of ring opening of *cis*-1,2-dimethyl-3-cyclopropyl cation.

Answers to Problems

Chapter 1

1. Start from cyclooctatetraene. It will react with chlorocarbene. This yields the bicyclic compound (A) which ring opens on treatment with lithium.

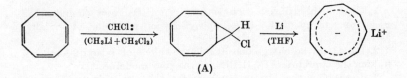

(A)

2. This also involves a carbene reaction:

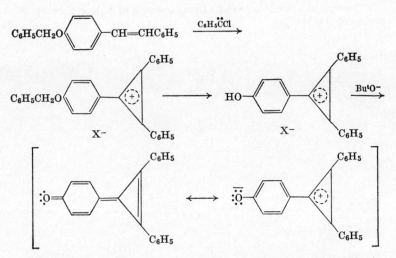

Not in fact very stable—suggesting that there is little resonance stabilization

3. The important fact here is that 2-hydroxybiphenylene couples in the 3-position. Subsequent reactions are quite normal. The quinone is very stable for an *o*-quinone (see Part 2).

4.

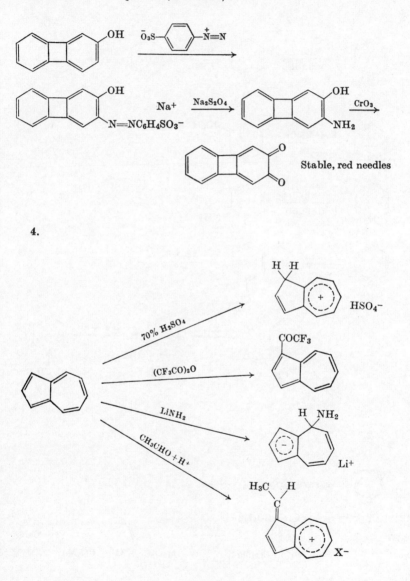

Stable, red needles

Chapter 2

1.

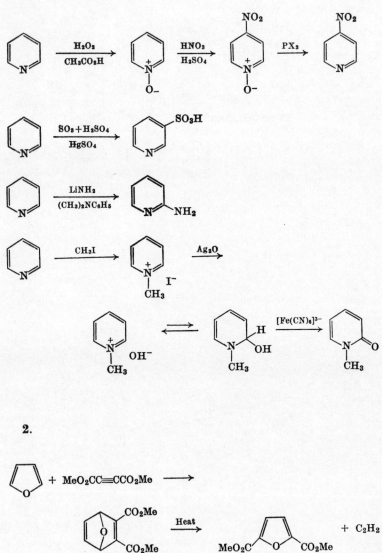

2.

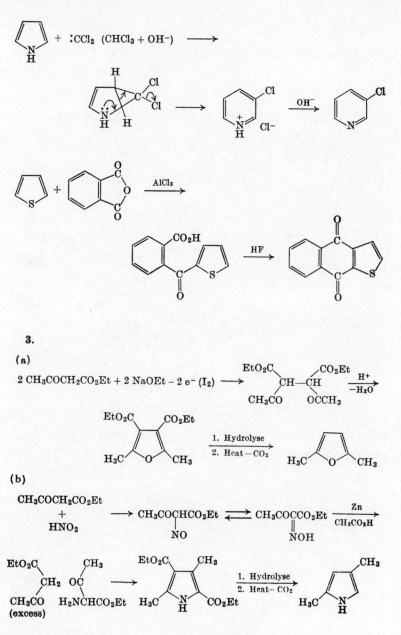

3.

(a)

$$2\ CH_3COCH_2CO_2Et + 2\ NaOEt - 2\ e^- \ (I_2) \longrightarrow$$

(b)

$$CH_3COCH_2CO_2Et + HNO_2 \longrightarrow$$

(c)

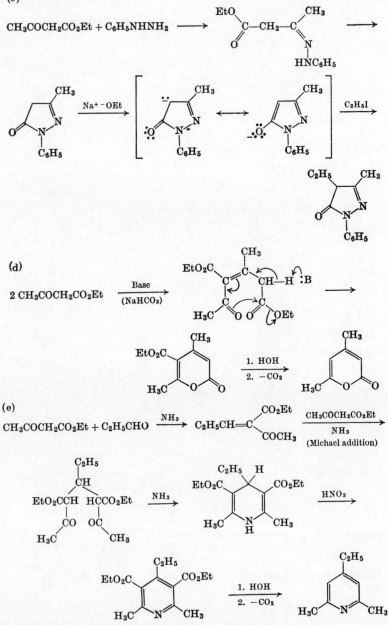

$CH_3COCH_2CO_2Et + C_6H_5NHNH_2 \longrightarrow$

(d)

$2\ CH_3COCH_2CO_2Et \xrightarrow[\text{(NaHCO}_3)]{\text{Base}}$

(e)

$CH_3COCH_2CO_2Et + C_2H_5CHO \xrightarrow{NH_3}$

(f)

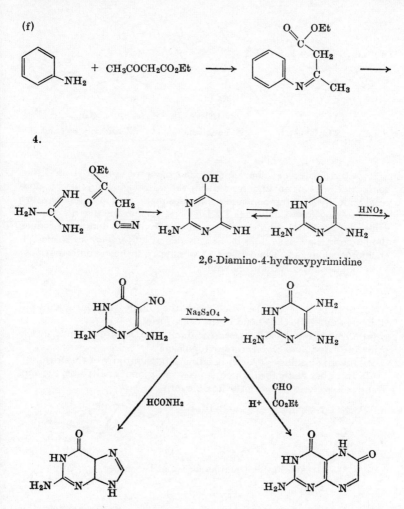

4.

2,6-Diamino-4-hydroxypyrimidine

Chapter 3

1. (A) = (B) = (C) = (+)-*cis*-form. (D) = (E) = the enantiomeric (−)-*cis*-form. (F) is *trans*-1-bromo-2-methylcyclohexane.

2.

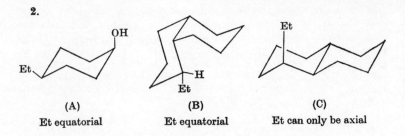

(A)	(B)	(C)
Et equatorial	Et equatorial	Et can only be axial

3. *cis*-1,2-Dichlorocyclohexane has no plane of symmetry in either conformer but ring inversion produces the mirror image. It is therefore resolvable at low temperatures but racemizes at room temperature owing to ring inversion. *trans*-1,2-Dichlorocyclohexane has no plane of symmetry in either conformation, and ring inversion does not produce the mirror image. Therefore it is resolvable at all temperatures. *cis*-1,3-Dichlorocyclohexane has a plane of symmetry in both conformations; therefore it is *meso* and unresolvable.

4. (*a*) There is repulsion between the dipoles of the equatorial C–Br bond and the C=O bond in the other form.

(*b*) There is repulsion between the dipoles of the C–Br bonds in the other form and possibly also steric repulsion.

(*c*) Intramolecular hydrogen-bonding is possible only in this form.

(*d*) In either chair form there is necessarily one axial t-butyl group. There is less steric strain in the flexible form.

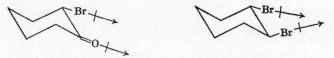

Alternative conformations of higher energy

5. In each case departure of the bromide ion is assisted by participation of the atom antiperiplanar to it.

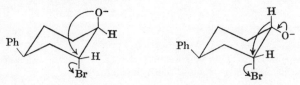

6. The $-CO_2^-$ group is more solvated and therefore more bulky than $-CO_2H$. When ionization occurs, the increase in bulk is better accommodated when the group is equatorial than when it is axial. In both isomers the t-butyl group will be equatorial. The *trans*-acid, in which the carboxy group is also equatorial, therefore forms an anion more readily than the *cis*-acid in which the carboxy group is necessarily axial.

7. Approach of the bromine molecule to the less hindered face is followed by *trans*-diaxial opening of the bromonium ion to give a chair ring. Note that the positive species, Br^+, becomes attached finally to the more substituted end of the double bond, contrary to what would be predicted on the basis of inductive effects (cf. page 176).

8. The *trans*-isomer is more stable by 2.9 kJ mole^{-1} (the equatorial preference of the OH). For *trans* $\rightleftarrows$ *cis*, $\log K = \dfrac{2900}{2.3 \times 8.31 \times 400}$, whence $K = 2.4$. The proportion of *trans* therefore is $\dfrac{2.4}{3.4} = 71\%$.

9. (*i*) (*a*) one; (*c*) not active

(*ii*) (*a*) two; (*b*) the *cis*-fused one; (*c*) neither

(*iii*) (*a*) five; (*b*) the *trans-syn-trans*; (*c*) the *cis-anti-trans* and *trans-anti-trans*

(*iv*) (*a*) one; (*c*) it is resolvable

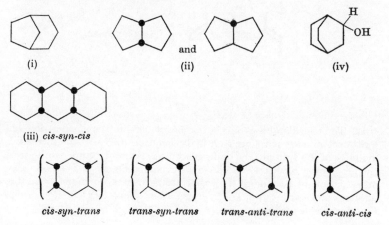

(i)

(ii) and

(iv)

(iii) *cis-syn-cis*

cis-syn-trans   *trans-syn-trans*   *trans-anti-trans*   *cis-anti-cis*

10. In the first case epimerization relieves severe 1,3-diaxial inter-actions. In the second case the epimer would have a strained boat ring.

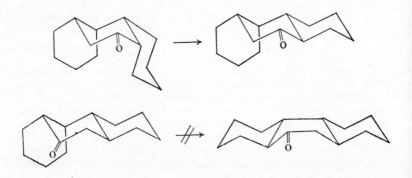

11. The two conformations of cyclooctene are enantiomeric and non-superposable, and their interconversion by ring inversion has a high activation energy. Ring inversion in *trans*-cyclodecene is easy owing to the greater mobility in this larger ring.

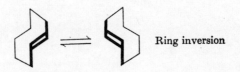

Ring inversion

12. The considerable strain in *trans*-cyclooctene (the smallest known *trans*-cycloalkene) is relieved on hydrogenation owing to the increased mobility of the system. The hydrogenation is therefore more exothermic than for simple olefins. However, in *cis*-cyclodecene, which is mobile, the introduction of two more hydrogen atoms causes an increase in the torsional and steric strain. The hydrogenation is therefore less exothermic than for simple olefins.

13. Transannular participation by the double bond gives the electronically preferred tertiary carbonium ion, the angle strain in which is relieved by bond migration.

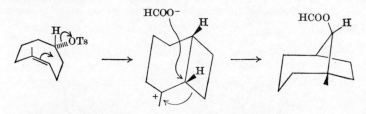

14. The ORD trough is at longer wavelength than the peak, therefore the Cotton effect is negative. Assuming that the ring is a chair, the diaxial conformer should show a negative Cotton effect and the diequatorial form a weak positive one. (The mass, polarizability, and proximity of the chlorine atom make it by far the more important factor.) In octane, therefore, this ketone adopts predominantly the diaxial conformation. Presumably in this non-polar solvent steric factors are outweighed by the reduction in energy to be gained by orienting the carbonyl and chlorine dipoles in different directions. In methanol as solvent the diequatorial conformation is adopted since dipole–dipole repulsion is reduced in this polar medium.

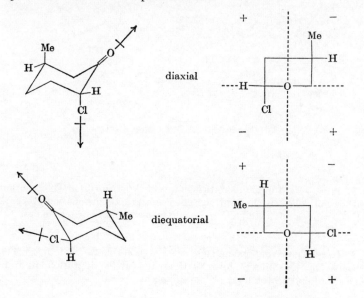

Chapter 4

1. In only this conformation is hydrogen-bonding possible. The reduction in energy resulting from this bonding more than compensates for any energy required to change the hydroxy group from equatorial to axial.

2. In the cyclohexane analogue (A), the equatorial preference of Bu^t is much greater than that of methyl owing to the greater steric interaction with the *syn*-axial hydrogens. In the dioxan if the methyl is axial (C), interaction with the *syn*-axial hydrogens is still present (and possibly more severe than in the cyclohexane owing to the shorter C–O bonds). If the t-butyl is axial, however, there are no *syn*-axial hydrogens (D) and, presumably, interaction with the lone pairs on oxygen is not strongly unfavourable and may possibly be favourable.

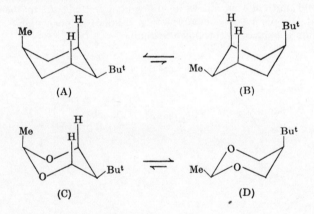

3. Both 5- and 6-membered rings are fairly strain-free. The reaction path is controlled by the inductive effect of the alkyl groups and proceeds via the more substituted and more stable carbonium ion is each case.

4. In each case ring opening gives the more stable carbonium ion. The reasons for the different fates of these ions (acyl migration, hydride migration, and proton loss) are not apparent.

(a)

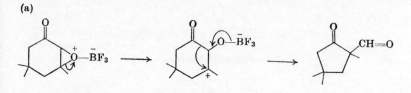

(b)

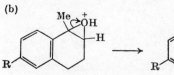

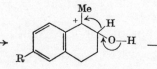

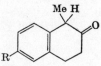

(c)

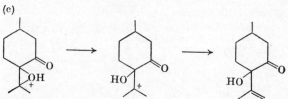

Chapter 5

1.

(a)

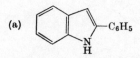

(d)

(b)

$$\begin{array}{c} CH_3O \\ \diagdown \\ CH_3O \end{array} P{=}O \quad + CH_3Br \\ \diagup \\ C_2H_5$$

(e) $(CH_3)_2S$

(c) $CCl_2{=}CCl{-}N(CH_3)_2$

(f)

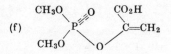

2.

(a) 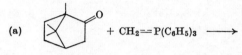 $+ CH_2{=}P(C_6H_5)_3 \longrightarrow$

(b) $C_6H_5CH{=}CH{-}CH_2Cl + (C_6H_5)_3P \xrightarrow[\text{reflux}]{\text{Xylene}}$

$\qquad C_6H_5CH{=}CH{-}CH_2\overset{+}{P}(C_6H_5)_3 \quad Cl^- \xrightarrow{\text{LiOC}_2\text{H}_5}$

$\qquad\qquad C_6H_5CH{=}CH{-}CH{=}P(C_6H_5)_3 \xrightarrow{C_6H_5CHO}$

(c) $CH_3CHBrCO_2CH_3 + (C_6H_5)_3P \xrightarrow[\text{reflux}]{\text{Benzene}}$

$\qquad\qquad \overset{\displaystyle CH_3}{(C_6H_5)_3\overset{+}{P}CHCO_2CH_3} \quad Cl^- \xrightarrow{\text{NaOH}}$

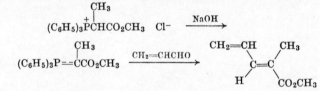

(d) $C_6H_5COCH_2Br + (C_6H_5)_3P \longrightarrow$

$\qquad\qquad C_6H_5COCH{=}P(C_6H_5)_3 \xrightarrow{(C_2H_5)_2CO}$

(e) $(CH_3)_3S^+ \ Br^- \xrightarrow{CH_3Li} (CH_3)_2S{=}CH_2 \xrightarrow{\text{Cyclo-C}_7H_{12}O}$

3

(a)

(b)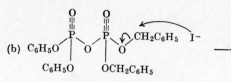

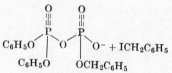

Chapter 6

1. (a) Rate $= k[C_2H_5Cl][KI] = k[0.1][0.1] = 5.44 \times 10^{-7}$

Therefore, $k = 5.44 \times 10^{-5}$

At 0.01M, Rate $= 5.44 \times 10^{-5}[0.01][0.01]$

$= 5.44 \times 10^{-9}$ mole l.$^{-1}$ sec^{-1}

(b) Rate $= k[C_2H_5Cl][KI]^2$

Therefore, $k = \dfrac{5.44 \times 10^{-7}}{(0.1)^3} = 5.44 \times 10^{-4}$

At 0.01M, Rate $= 5.44 \times 10^{-4} \times (0.01)^3$

$= 5.44 \times 10^{-10}$ mole l.$^{-1}$ sec^{-1}

14

(c)

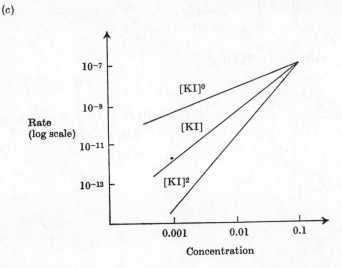

(d) If the concentration of just the potassium iodide were changed, the rate would remain unchanged if it were zero-order in potassium iodide, but it would change markedly ($\propto [KI]$) if first-order in potassium iodide.

2. (a) [K. A. Saegebarth, *J. Org. Chem.*, **25**, 2212 (1960)]

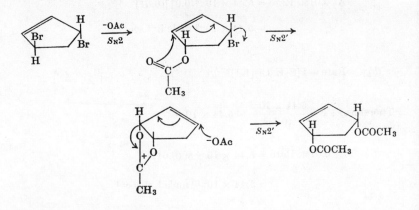

(b) [R. Heck and S. Winstein, *J. Amer. Chem. Soc.*, **79**, 3105 (1957)]

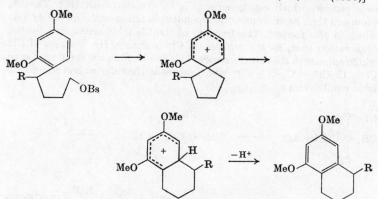

(c) [D. S. Noyce and H. I. Weingarten, *J. Amer. Chem. Soc.*, **79**, 3093 (1957)]

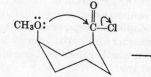

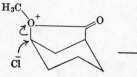

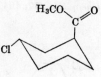

(d)

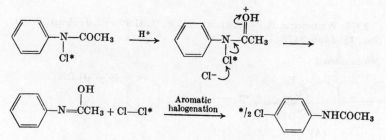

(contd.)

This is the Orton rearrangement. The *N*-chloro-amide merely serves as a reagent which can generate Cl_2 by reaction with HCl. The Cl_2 produced then carries out an independent reaction with the acetanilide which is also formed. The freedom of the Cl_2 is demonstrated by the observation that, as the amount of Cl^- increases, the fraction of Cl^* which appears in the ring decreases. This results from the rapid exchange $Cl^- + Cl-Cl^* \rightleftharpoons Cl-Cl + Cl^{*-}$ which causes all the chlorine and chloride to be in equilibrium.

(e)

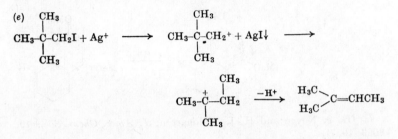

Ag^+ assists in the ionization of the alkyl iodide (electrophilic catalysis) to give a carbonium ion which can undergo rearrangement. The mechanism given undoubtedly represents an oversimplification, since it may be questioned whether a full carbonium ion is generated before the rearrangement begins, or whether the loss of halide ion is assisted by the movement of the methyl group.

(*f*) [H. L. Goering, T. D. Newitt, and E. F. Silversmith, *J. Amer. Chem. Soc.*, **77**, 4042 (1955)]

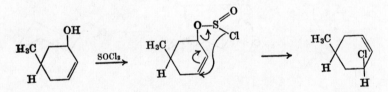

3. [J. Weinstock, R. G. Pearson, and F. G. Bordwell, *J. Amer. Chem. Soc.*, **78**, 3468, 3473 (1956)]

Mechanism 1

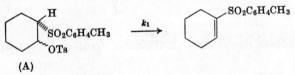

(A)

Mechanism 2

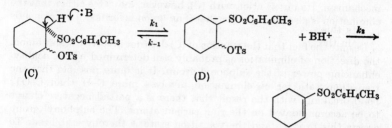

(C)   (D)

Two possible mechanisms for elimination reactions are shown above. The first, concerted one (*E*2) is that normally found for cases in which the *trans* leaving groups and the carbon atoms connecting them can be placed in one plane. It is presumably applicable in the case of compound (A). The *cis*-elimination from (C) is more surprising since there are hydrogens available for *trans*-elimination as well. It was originally suggested that the electron-withdrawing effect of the sulphonyl group enhanced the acidity of the proton adjacent to it, and that the reaction proceeded by the formation of an anion which subsequently lost a tosyl ion (mechanism 2). Two cases should be considered with regard to such a mechanism. In the first, $k_{-1} > k_2$, and an equilibrium concentration of the anion is formed. In the second, $k_2 > k_{-1}$.

The two cases may be distinguished by the results with amine catalysis.

Case 1. The rate of disappearance of (C) is given by

$$\text{Rate} = k_2[D]$$

and since at equilibrium, $k_{-1}[D][BH^+] = k_1[C][B]$

$$[D] = \frac{k_1[C][B]}{k_{-1}[BH^+]}$$

Hence, $\text{Rate} = k_2 \dfrac{k_1[C][B]}{k_{-1}[BH^+]}$.

Case 2. $\text{Rate} = k_1[C][B]$

In case 1, the observed rate is determined by the ratio $[B]/[BH^+]$ and not by the concentration of B alone. Since this same ratio determines the pH of the buffer, this behaviour is often referred to as specific OH^- catalysis.

In case 2, however, there is no return from the anion, and every base in the medium which can help to remove a proton contributes its share to the observed rate, regardless of the concentration of its conjugate acid. This is general base catalysis.

The increase in rate with increasing amine concentration observed with (A) is reasonable, since no return is expected for the concerted mechanism. Its observation with (C), however, indicates either that the elimination is also concerted or that if an anion is formed it corresponds to case 2.

Despite the fact that the reaction of (C) does not involve a stable anion, the direction of elimination is probably still determined by the acidity-enhancing power of the sulphonyl group. It is quite possible that the transition state for *cis*-elimination involves more C–H than C–OTs bond-breaking, with the result that there is a partial negative charge to be accommodated on the ring carbon atoms. The sulphonyl group assists this process, and the transition state is thereby stabilized. To explain the results, this stabilization must be enough to overcome the usual energy advantage for the *trans* process. The greater rate for elimination from (A) shows the increased advantages of having both effects operative.

4. When R = H, $k_2 = 42 \times 10^{-5}$ l mole^{-1} sec^{-1}

$$\rho = +2.14$$

The fact that ρ is positive means that the reaction develops some carbanion character but at the transition state this is more pronounced for the elimination of HF than for the elimination of HBr (this is revealed by the fact that the HF elimination has a large positive ρ, meaning that its transition state can be stabilized to a greater extent by groups which would stabilize a carbanion).

Consider the transition state:

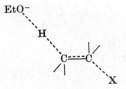

As ethoxide ion begins to remove a proton from the bromide (X = Br) some carbanion character develops, but bromide is such a good leaving group that it quickly begins to depart and the new double bond starts forming. Consequently, only a moderate anionic charge ever develops on the carbon; this means that the substituent effects are limited in size. When fluoride ion is the leaving group (X = F) more negative charge must develop before this ion can be ejected (fluoride is a poor leaving group); this makes the reaction more sensitive to substituent effects.

5. (*a*) Shift from Saytzeff-type orientation to Hofmann-type orientation as the degree of steric crowding is increased.

(*b*) S_N1 reaction cannot occur because the formation of a planar carbonium ion is prohibited by the rigidity of the bicyclic system.

S_N2 reaction is virtually impossible because of the inaccessibility of C-1 to back-side attack. In addition, the bonds to this carbon atom are held rigid, therefore Walden inversion about such a carbon atom cannot occur.

Elimination reactions are out of the question because of the impossibility of forming a double bond at the bridgehead of a bicyclo[2,2,1]heptane.

(*c*) The rate of incorporation of radioactive bromide is a measure of the rate at which bromide ion undergoes nucleophilic substitution with the alkyl bromide. If there is complete inversion in each substitution, the rate of substitution will be one-half the rate of loss of optical activity because one-half mole of inverted product will just cancel the rotation of one-half mole of starting material. Therefore the substitution of Br by Br* proceeds with clean inversion.

(*d*) In the first case the reaction will proceed by an S_N2 mechanism, and the large steric effect of the neopentyl group compared to that of the ethyl group will explain the rate ratio.

In the second case the mechanism can change to an S_N2' mechanism when R will not have such a profound steric effect on the relative rates of substitution.

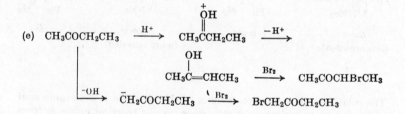

(*e*)

In basic solution the secondary enolate anion is expected to be less stable, and hence formed less rapidly, than the primary enolate anion because of the destabilizing effect of inductive electron-release by the methyl substituent in the secondary enolate anion.

In acid solution it is expected that the enol with the more highly substituted double bond will be more stable and therefore formed preferentially.

(*f*) [S. Winstein, H. V. Hess, and R. E. Buckle, *J. Amer. Chem. Soc.*, **64**, 2796 (1942)]

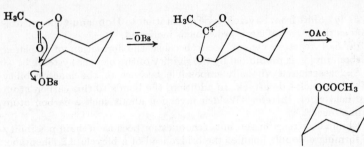

(g) [D. J. Cram, F. D. Greene, and C. H. Depuy, *J. Amer. Chem. Soc.*, **78**, 790 (1956)]

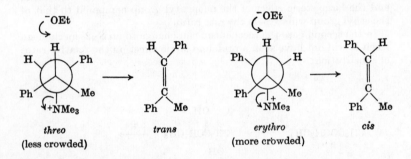

The relatively slow reaction leading to the *cis*-olefin is sometimes said to be due to an 'eclipsing effect' (since, as the reaction proceeds from the transition state to the olefinic product, one phenyl group appears to 'pass in front of' the other).

(h) The low C–Br bond energy compared with that of the C–Cl bond facilitates ionization of the bromide and causes it to react more rapidly than the chloride. However, since both reactions involve the formation of the same carbonium ion and the behaviour of this carbonium ion is independent of the nature of the leaving group, then the ratio of olefin to alcohol in the mixture of products will be the same for both halides.

(i) Addition of Li_2SO_4 accelerates the hydrolysis of benzhydryl chloride by the special salt effect.

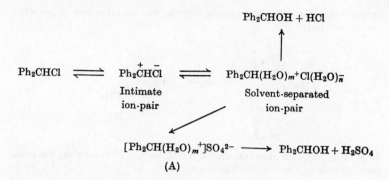

In the above scheme, SO_4^{2-} can intervene after the second step, converting some of the solvent-separated ionic chloride into a solvent-separated sulphate (A) which is very rapidly converted into solvolysis products. Intervention of sulphate lessens the likelihood of reversal of the second step, thus increasing the overall rate of solvolysis. Azide ion, on the other hand, is a much more effective nucleophile than sulphate anion, and is therefore much more effective in diverting the benzhydryl group to benzhydryl azide.

(*j*) [F. H. Westheimer and N. Nicolaides, *J. Amer. Chem. Soc.*, **71**, 25 (1949)]

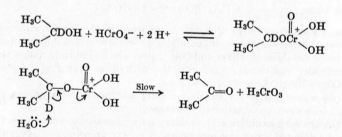

(*k*) [S. Cristol and W. P. Norris, *J. Amer. Chem. Soc.*, **76**, 3005 (1954)] The data can be explained in terms of a concerted process for *trans*-elimination and a multiple-stage carbanion-intermediate process for *cis*-elimination.

(*l*) The function of an acid and a base in catalysing the enolization of acetone (i.e. the rate-determining step in the bromination of acetone) can be written:

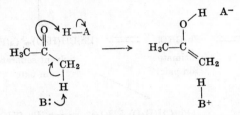

In a buffer solution of acetic acid and sodium acetate in water, the complete kinetic expression for this reaction is:

$$\text{Rate} = k_0[\text{Acetone}] + k_1[\text{Acetone}][\text{H}_3\text{O}^+] + k_2[\text{Acetone}][\text{HOAc}]$$
$$+ k_3[\text{Acetone}][\text{OH}^-] + k_4[\text{Acetone}][\text{OAc}^-]$$
$$+ k_5[\text{Acetone}][\text{HOAc}][\text{OAc}^-]$$

The first term represents enolization accomplished simply by water; the second term represents enolization in which H_3O^+ is the proton donor and H_2O the base, etc.

Terms in which OAc^- and H_3O^+ or HOAc and OH^- cooperate are already hidden in the expression we have written.

For example, in water, acetic acid is in equilibrium with H_3O^+ and OAc^-

$$\text{HOAc} \underset{\longleftarrow}{\overset{K}{\longrightarrow}} \text{H}_3\text{O}^+ + \text{OAc}^-$$

$$[\text{H}_3\text{O}^+][\text{OAc}^-] = K[\text{HOAc}]$$

Therefore, $$k[\text{H}_3\text{O}^+][\text{OAc}^-] = k'[\text{HOAc}]$$

This equilibrium shows that the third term in the kinetic expression does not necessarily represent catalysis by acetic acid, but it may actually indicate catalysis by H_3O^+ and OAc^-, since the equilibrium makes them kinetically equivalent. Similarly, the fifth term may really indicate catalysis by HOAc and OH^-.

(*m*) σ is negative for electron donors and positive for groups that withdraw electrons relative to hydrogen.

Groups such as chloro and methoxy have unshared electron pairs that can be donated by a resonance effect. However, chlorine and oxygen are more electronegative than carbon or hydrogen, and the inductive effect of both of these atoms is in the opposite direction; that is, they withdraw electrons by their inductive effects, but they can donate electrons by resonance; conjugation is, of course, required for the latter effect to operate. In the case of methoxy the σ_{para} value is negative, indicating that the resonance effect is greater than the inductive effect, but σ_{meta} is positive (inductive effect only, no conjugation present). In the case of chlorine the σ_{para} value is positive, indicating that the inductive effect is greater than the resonance effect. The σ_{meta} value is also positive, but larger, indicating that the only effect is the inductive effect.

(*n*) The transition state for the addition of bromine to *para*-substituted styrenes, namely (A):

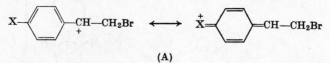

(A)

involves direct resonance of the substituent X, with a positive centre. In such a case the Hammett correlation holds only if σ^+ parameters are used.

6. By using 'crossover experiments'. Carry out the reaction with a mixture of two similar but non-identical reactants, and search the product for compounds having fragments of both reactants, thus seeing whether fragments from one reactant have 'crossed over' and have become attached to fragments of the other reactant.

7. The Curtius rearrangement of benzoyl azide (B; R = H) proceeds as shown below.

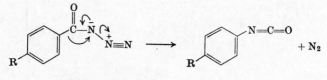

(B)

If the four-membered ring compound (A) was an intermediate in this reaction then a *para*-substituted benzoyl azide would yield a *meta*-substituted phenyl isocyanate.

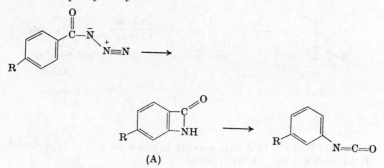

(A)

8. Strong evidence for the participation of the cyclic ion (B) in the reaction would come from the stereochemistry of the product (C). Thus:

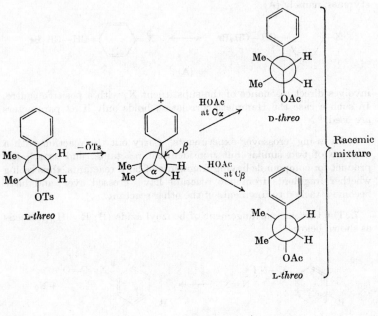

but

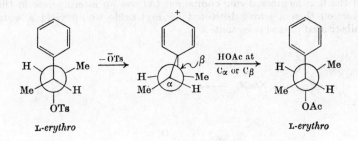

9. [J. D. Roberts, D. A. Semenow, H. E. Simmons, and L. A. Carlsmith, *J. Amer. Chem. Soc.*, **78**, 601 (1956)]

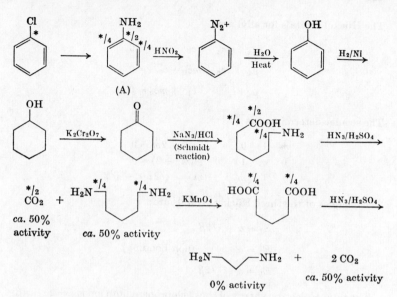

(A)

The distribution of ^{14}C in the product, aniline, is expected to be as shown in (A). Radioactivity measurements of the degradation products in the reaction have clearly shown the validity of this assumption.

Chapter 7

1. The Hückel orbitals for cyclopropyl are:

$$\mu_a \qquad mc_a + \quad c_b + c_c \quad = 0$$
$$\mu_b \qquad c_a + mc_b + c_c \quad = 0$$
$$\mu_c \qquad c_a + \quad c_b + mc_c = 0$$

The secular determinant is:

$$\begin{vmatrix} m & 1 & 1 \\ 1 & m & 1 \\ 1 & 1 & m \end{vmatrix} = 0$$

Therefore, $m^3 - 3m + 2 = 0$
$$(m - 1)^2(m + 2) = 0$$

Thus the three molecular orbitals have energies:

$$\psi_1 \quad E_1 = \alpha + 2\beta$$

ψ_2 and $\psi_3 \quad E_2 = E_3 = \alpha - \beta$ (degenerate orbitals)
In the cation two electrons go into E_1

The Hückel orbitals for allyl are:

$$\mu_a \quad mc_a + c_b \qquad = 0$$

$$\mu_b \qquad c_a + mc_b + c_c = 0$$

$$\mu_c \qquad\qquad c_b + mc_c = 0$$

The secular determinant is:

$$\begin{vmatrix} m & 1 & 0 \\ 1 & m & 1 \\ 0 & 1 & m \end{vmatrix} = 0 \qquad \begin{aligned} m^3 - 2m &= 0 \\ m(m^2 - 2) &= 0 \\ m = -\sqrt{2}&; \, 0, \, +\sqrt{2} \end{aligned}$$

The energies of the three Hückel orbitals are:

$$E_1 = \alpha + \sqrt{2}\beta$$

$$E_2 = \alpha \qquad \text{(non-bonding)}$$

$$E_3 = \alpha - \sqrt{2}\beta$$

The total π-electron energy of the cyclopropenylium ion is $\epsilon = 2\alpha + 4\beta$, which is lower than the total π-electron energy of the allyl cation $\epsilon = 2\alpha + 2\sqrt{2}\beta$.

2.

The starred atoms in the two molecules are:

Let the coefficients of the NBO be a:

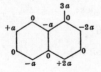 

Normalizing:

$$20a^2 = 1$$
$$a = \frac{1}{\sqrt{20}}$$

$$17a^2 = 1$$
$$a = \frac{1}{\sqrt{17}}$$

Thus the charges on the anion are:

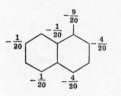 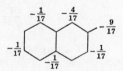

(Cf. Part 2, p. 262)

3.

Dewar numbers for phenanthrene

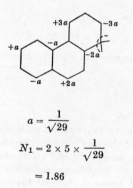 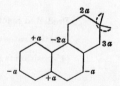 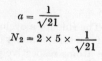

$$a = \frac{1}{\sqrt{29}}$$
$$N_1 = 2 \times 5 \times \frac{1}{\sqrt{29}}$$
$$= 1.86$$

$$a = \frac{1}{\sqrt{21}}$$
$$N_2 = 2 \times 5 \times \frac{1}{\sqrt{21}}$$
$$= 2.18$$

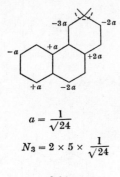

$$a = \frac{1}{\sqrt{24}}$$

$$N_3 = 2 \times 5 \times \frac{1}{\sqrt{24}}$$

$$= 2.04$$

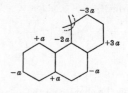

$$a = \frac{1}{\sqrt{26}}$$

$$N_4 = 2 \times 5 \times \frac{1}{\sqrt{26}}$$

$$= 1.96$$

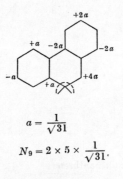

$$a = \frac{1}{\sqrt{31}}$$

$$N_9 = 2 \times 5 \times \frac{1}{\sqrt{31}}.$$

$$= 1.80$$

$\therefore$ Predicted reactivity $9° > 1° > 4° > 3° > 2°$

Relative rates of nitration $9° > 1° > 3° > 2° > 4°$

($4°$-position is subject to considerable steric hindrance)

4. The rule for a cycloaddition between two neutral molecules is that if half the total number of atoms (q) of the two reactants (made up of m and n atoms respectively) is odd (i.e. $m + n = q$ and $q/2$ is odd) then a thermal reaction is expected. For two butadiene molecules we have $4 + 4 = 8$, therefore $q/2$ is even, and hence a thermal reaction is not expected but a photochemical one should be possible. In practice a thermal reaction does occur under certain conditions but the product is vinylcyclohexene, which is in accord with orbital symmetry predictions

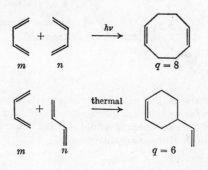

5. We deal with this molecule by treating the orbital on the trivalent carbon atom as a p-atomic orbital (denoted ω). We now have a three-carbon fragment, and our rules applied only to even numbered systems. However, the pairing properties of Hückel orbitals remain the same, so that we can take note of the number of electrons, which will tell us the number of occupied orbitals. In the anion there will be two electrons in the ω-orbital and two in the σ-orbital being broken. Half the number of electrons is therefore even and we thus predict conrotatory ring opening for the anion

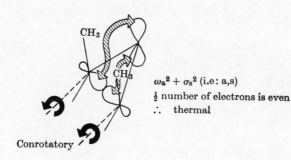

$\omega_a{}^2 + \sigma_s{}^2$ (i.e: a,s)
$\frac{1}{2}$ number of electrons is even
∴ thermal

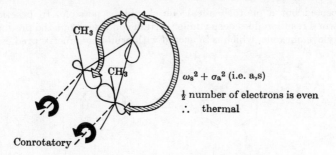

$$\omega_s{}^2 + \sigma_a{}^2 \text{ (i.e. a,s)}$$

$\frac{1}{2}$ number of electrons is even

$\therefore$ thermal

Notice that, for the cation with two electrons less, we would have a system where half the number of electrons was odd, and we would therefore predict that the ring opening would involve either an *s,s*-process or an *a,a*-process. Both of these are disrotatory.

Subject Index